Book 9S

PUBLISHED BY THE PRESS SYNDICATE OF THE UNIVERSITY OF CAMBRIDGE
The Pitt Building, Trumpington Street, Cambridge, United Kingdom

CAMBRIDGE UNIVERSITY PRESS
The Edinburgh Building, Cambridge CB2 2RU, UK
40 West 20th Street, New York, NY 10011-4211, USA
477 Williamstown Road, Port Melbourne, VIC 3207, Australia
Ruiz de Alarcón 13, 28014 Madrid, Spain
Dock House, The Waterfront, Cape Town 8001, South Africa

http://www.cambridge.org/

© The School Mathematics Project 2003
First published 2003

Printed in the United Kingdom at the University Press, Cambridge

Typeface Minion   *System* QuarkXPress®

A catalogue record for this book is available from the British Library

ISBN 0 521 53815 7 paperback

Typesetting and technical illustrations by The School Mathematics Project
Illustrations on pages 47 and 51 by Simon Hayes
and on pages 48 and 52 by Valerie Grace
Other illustrations by Robert Calow and Steve Lach at Eikon Illustration
Cover image © ImageState Ltd
Cover design by Angela Ashton

The publishers thank the following for supplying photographs:
Page 48 English Heritage Photographic Library
Pages 73, 89 and 90 Paul Scruton
All other photographs by Graham Portlock

NOTICE TO TEACHERS
It is illegal to reproduce any part of this work in material form (including photocopying and electronic storage) except under the following circumstances:
(i) when you are abiding by a licence granted to your school or institution by the Copyright Licensing Agency;
(ii) where no such licence exists, or where you wish to exceed the terms of a licence, and you have gained the written permission of Cambridge University Press;
(iii) where you are allowed to reproduce without permission under the provisions of Chapter 3 of the Copyright, Designs and Patents Act 1988.

Contents

1 Solids *4*
2 Equivalent expressions *13*
3 Fractions *20*
4 As time goes by *27*
5 Working with rules *37*
   Review 1 *44*
6 Circumference of a circle *46*
7 Clue-sharing *54*
8 Enlargement *55*
9 Over to you *61*
10 Straight-line graphs *63*
11 Points, lines and arcs *73*
12 Percentage problems *77*
   Review 2 *83*
13 Ratio and proportion *85*

14 Angles of a polygon *91*
15 Using and misusing statistics *99*
16 Linear sequences *107*
17 Decimals *115*
18 Area of a circle *127*
   Review 3 *132*
19 The right connections *135*
20 Algebra problems *144*
21 Angles *151*
22 Transformations *156*
23 Trial and improvement *167*
24 Spot the errors *174*
25 Stretchers *178*
   Review 4 *180*
   Index *183*

1 Solids

This work will help you
- identify and draw cross-sections of solids
- find the volume and surface area of a prism
- identify the planes of symmetry of a solid

A Cross-sections

A screwdriver is being lowered into some water.

The drawing for each photo shows the cross-section of the screwdriver at water level.

A1 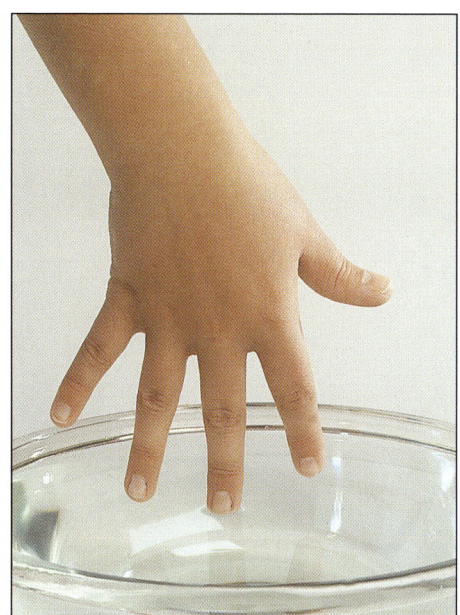 Nicky is lowering her hand into water. List these cross-sections of her hand in the right order.

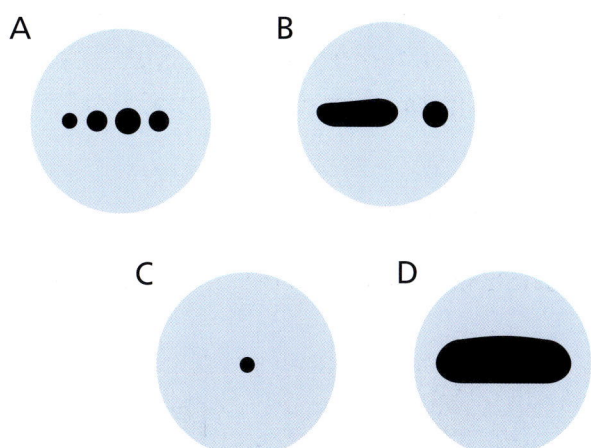

A2  This spanner is being lowered into water. List these cross-sections in the right order.

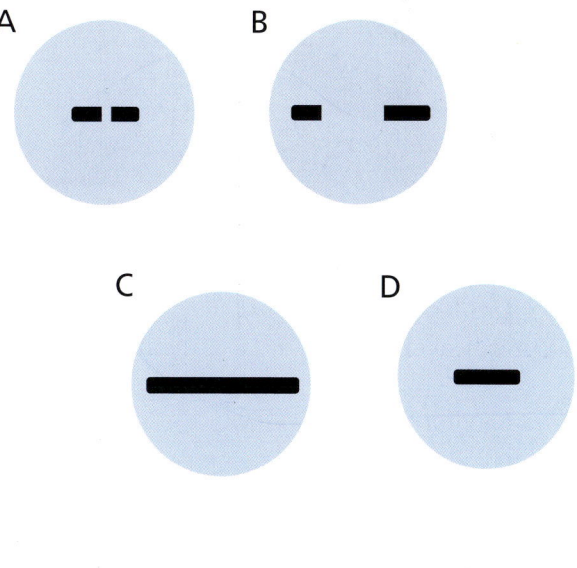

A3 Sketch a set of cross-sections for each of these.

Garden trowel

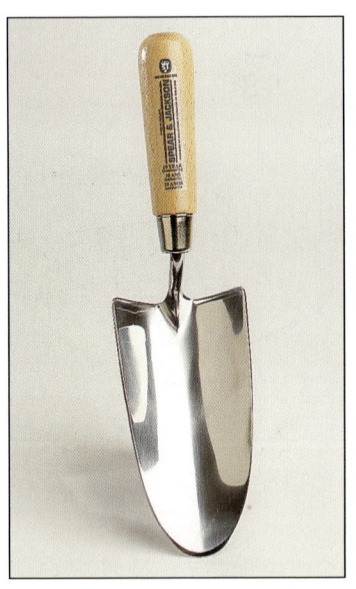

Garden hoe blade

Mallet

A4

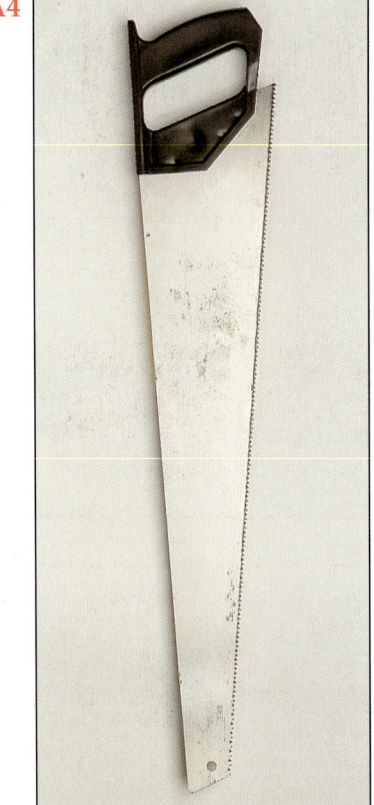

This saw is going to be lowered into water.
List these cross-sections in the right order.

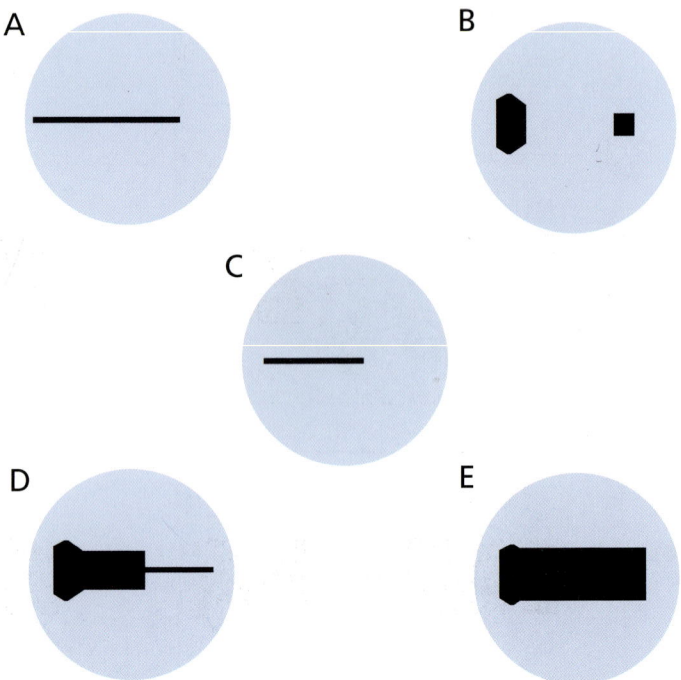

A5

This block is lowered into the water in three different ways.

1

2

3

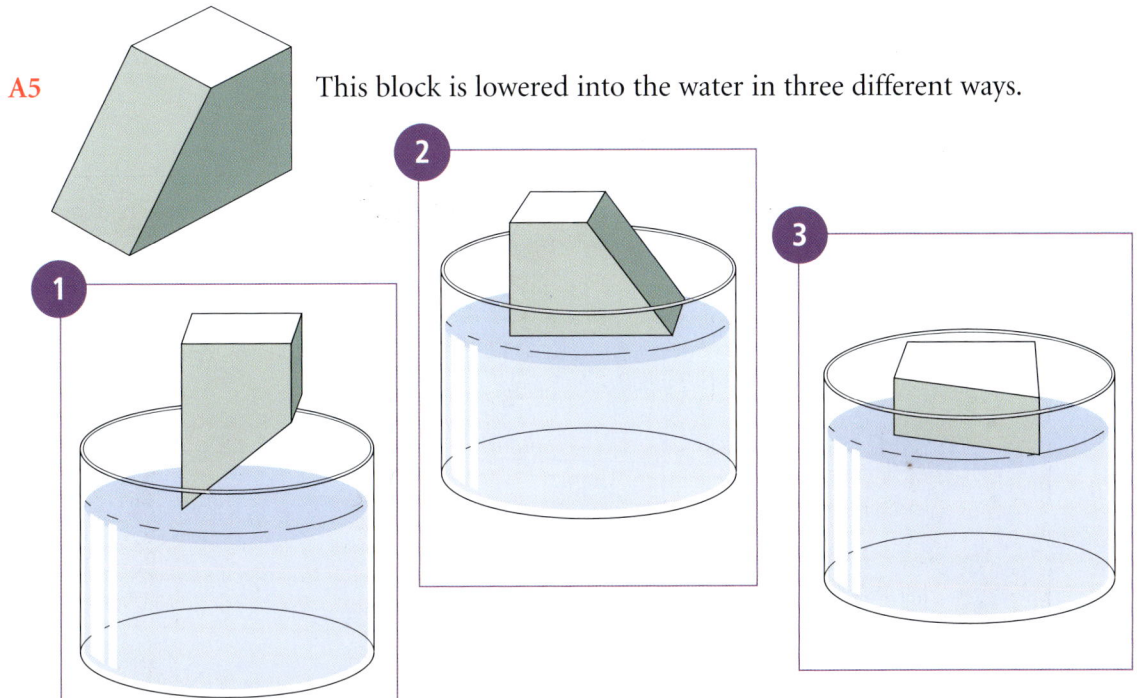

Match each set of cross-sections to the correct picture.

A

B

C

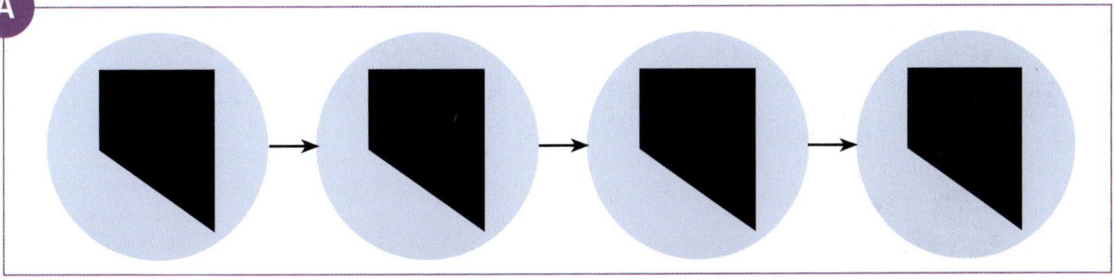

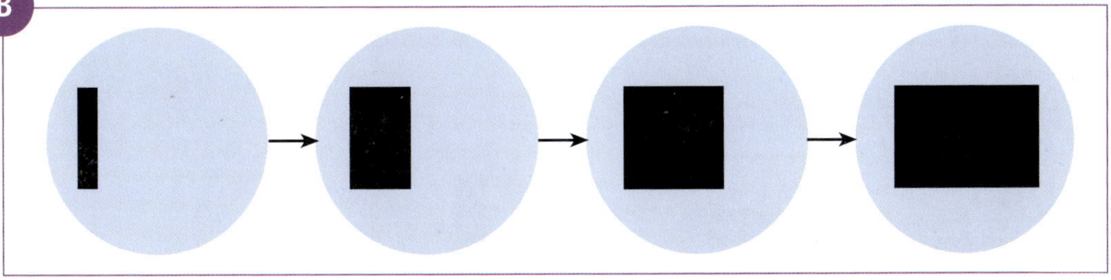

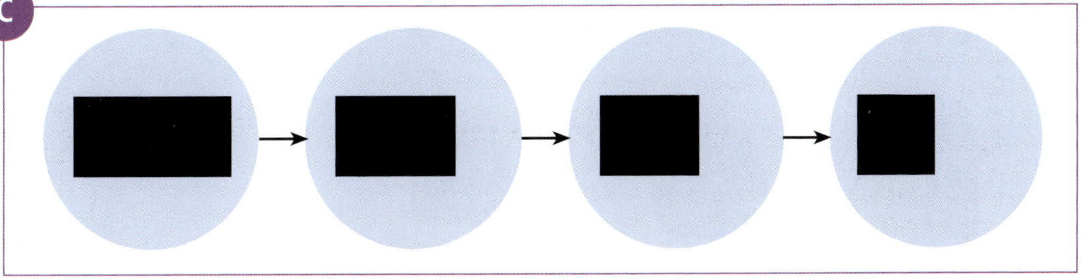

A6 This jar is being lowered into water in two different ways.

Sketch a set of cross-sections for each of the ways.

(a)    (b)

A7 Here are three different jars.

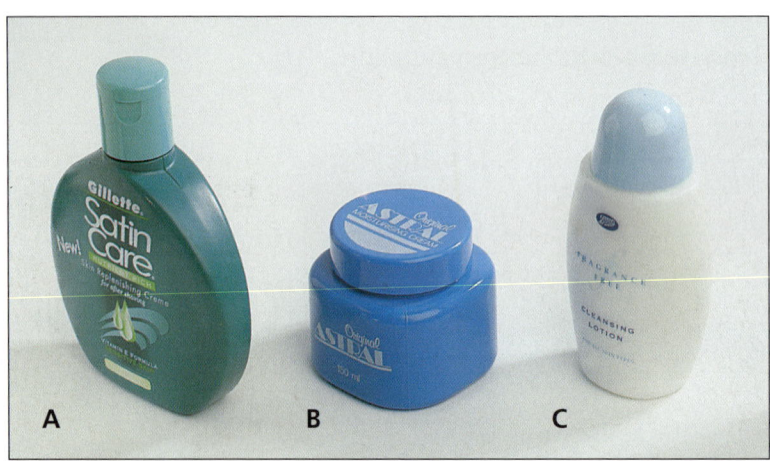

Which jar does each of these cross-sections belong to?

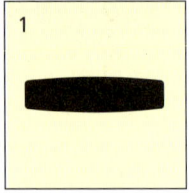 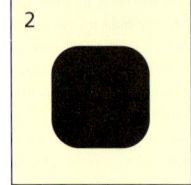 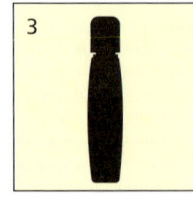

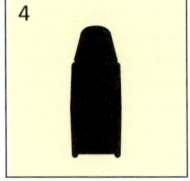 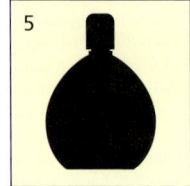 

A8 (a) What object do you think each of these sets of cross-sections shows?

(i)

(ii)

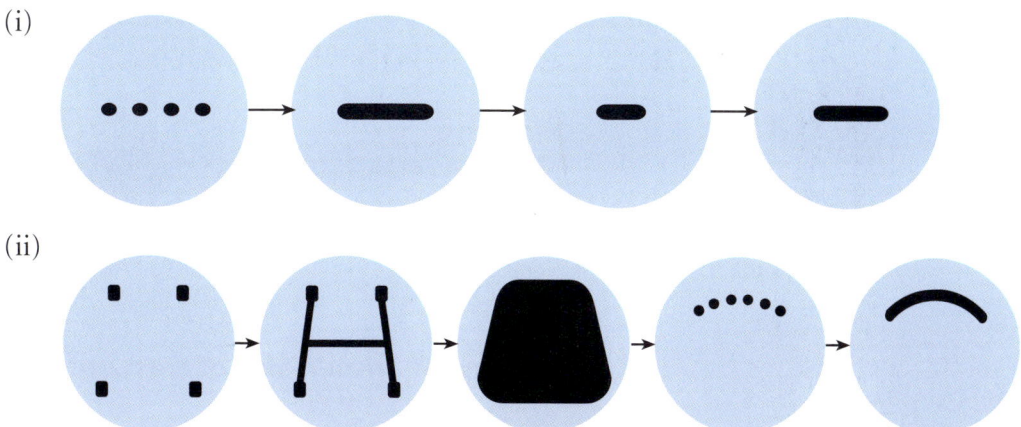

(b) Choose your own object and make up a similar puzzle.

B Prisms

These shapes are prisms…

… but these are not.

A prism with cross-sectional area 1 cm² and length 4 cm has a volume of 4 cm³.

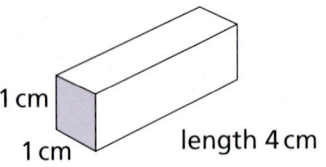

A shape made up of six such prisms will have a volume of
6 × 4 cm³ = 24 cm³.

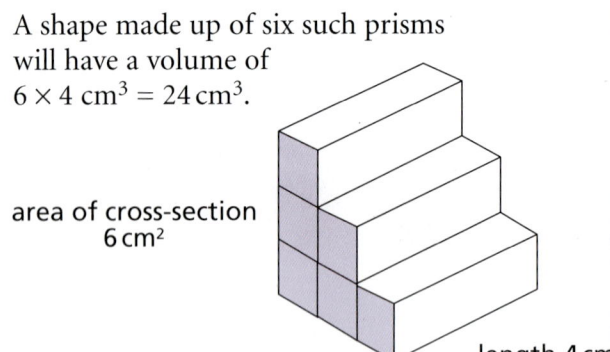

Volume of prism = area of cross-section × length

B1 Find the volume of each of these prisms.
They are all made from centimetre cubes (or parts of them).

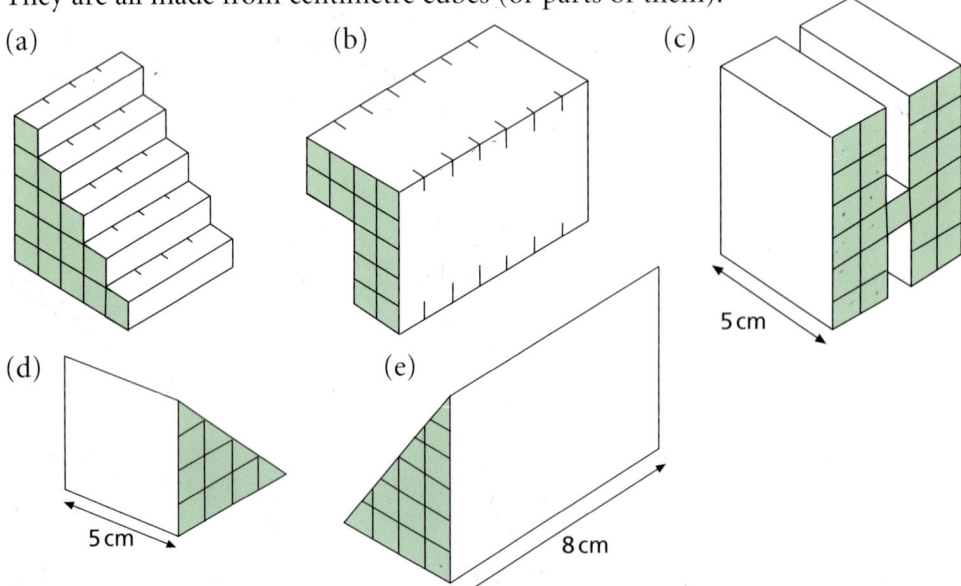

B2 Calculate the volume of each of these prisms.

B3 For each of the prisms in B2,
- make a sketch of each of its faces, marking in the dimensions
- calculate the area of each face
- add the areas to find the total surface area of the prism

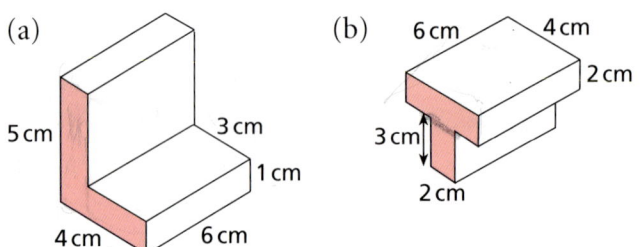

10

B4 Calculate the volume of each of these prisms.

(a) 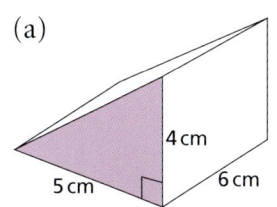 (b) 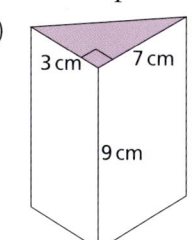 (c) 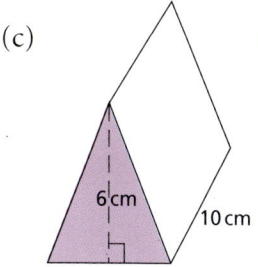 (d)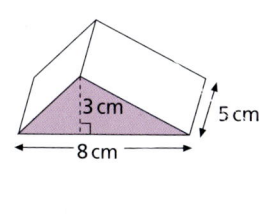

B5 Convert the lengths given in B4 to millimetres.
Calculate each volume in mm³. What do you notice?

C Planes of symmetry

The dotted line is a line of symmetry of this two-dimensional shape.

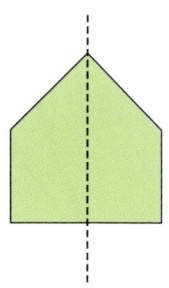

The shaded plane is a **plane of symmetry** of this three-dimensional shape.

This shape has another plane of symmetry.

Where is it?

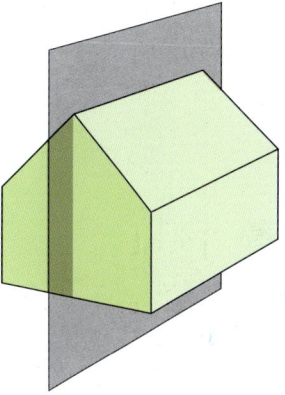

C1 The blue solid is a prism whose cross-section is an equilateral triangle.
One of its planes of symmetry is shown.

How many planes of symmetry does the prism have altogether?

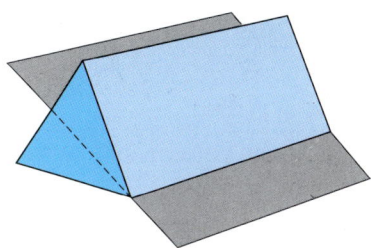

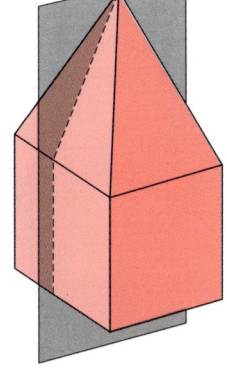

C2 The red solid consists of a square-based pyramid on top of a cube.
One of its planes of symmetry is shown.

How many planes of symmetry does the solid have altogether?

C3 How many planes of symmetry does this yellow cuboid have?

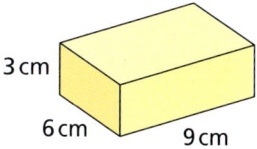

What progress have you made?

Statement

I can identify cross-sections of a solid.

Evidence

1 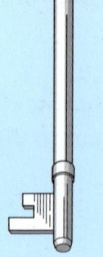 This key is lowered into water. List these cross-sections in the correct order.

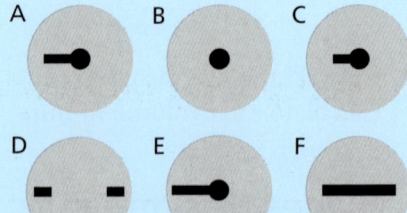

I can draw cross-sections of a solid.

2 Draw a set of four or five cross-sections of these as they are lowered into water.

(a) 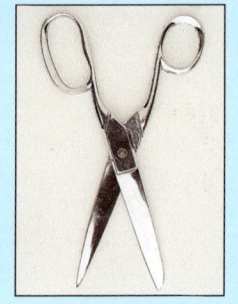 (b)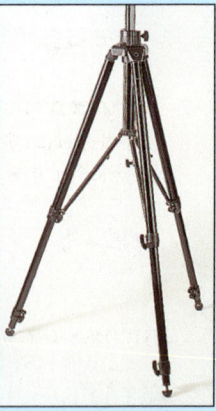

I can find the volume and surface area of a prism.

3 Find the volume and surface area of this prism.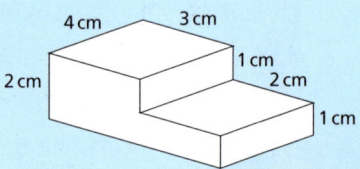

I can identify the planes of symmetry of a solid.

4 How many planes of symmetry does this square-based prism have?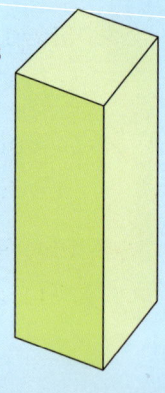

12

2 Equivalent expressions

This work will help you
- multiply out brackets in expressions such as $3(2n + 1)$, $n(n - 3)$
- factorise expressions such as $6n + 8$, $n^2 + 5n$
- simplify expressions that involve simple algebraic fractions

A Multiplying out brackets 1

	A	B	C	D	E	F	G
1	n	$2n+1$	$2n+2$	$2n+4$	$2(n+1)$	$2(n+2)$	$2(n+4)$
2	1	3	4	6	4	6	10
3	2	5	6				
4	0.5						

A1 The green shape is a square.
Which two of these expressions give the perimeter of the square?

$4(n + 3)$ $4n + 3$ $4(n + 12)$ $4n + 12$

The square has side $n + 3$.

The perimeter is the total distance round the edge of the shape.

A2 Multiply out the brackets in each of these.
(a) $2(n + 5)$ (b) $3(n - 4)$ (c) $6(n + 2)$ (d) $5(n - 3)$
(e) $4(n + 1)$ (f) $10(3 + n)$ (g) $2(7 - n)$ (h) $3(3 - n)$

A3 Hal has five flower pots and plants p seeds in each one.
All the seeds grow but a slug eats a plant from each pot.
Which two of these expressions give the total number of plants Hal has now?

$5p - 5$ $p + 5$ $5p - 1$ $5(p - 5)$ $5(p - 1)$ $p - 1$

A4 Simplify these expressions.
(a) $2 \times 3n$ (b) $3 \times 5n$ (c) $4 \times 4n$ (d) $5 \times 6n$

A5 Multiply out the brackets in each of these.
(a) $2(3n + 2)$ (b) $3(4n - 5)$ (c) $5(2n + 1)$ (d) $4(3n - 5)$

A6 Simplify these expressions.
(a) $3(n + 2) - 5$ (b) $2(n - 5) + 13$ (c) $2(3n + 4) - 3$

A7 (a) For the diagram below, what is the output when the input is 3?

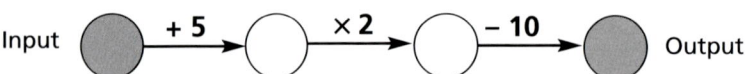

(b) Copy and complete this table for the arrow diagram above.

Input	Output
1	
4	
5	
10	

(c) Look at your completed table.
What simple rule links the inputs and outputs?

(d) (i) Copy and complete the arrow diagram below for an input of n.
Simplify the output expression as far as you can.

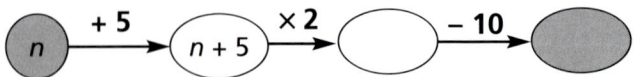

(ii) Explain how this shows the rule you found in (c) will work for **any** input.

A8 (a) For the diagram below, what is the output when the input is 5?

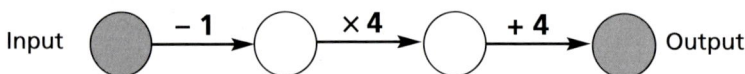

(b) (i) What expression is the output for an input of n?
Simplify this expression as far as you can.

(ii) Give the simple rule that links the inputs and outputs and explain how you know it will work for **any** input.

A9 (a) For the diagram below, what is the output when the input is 2?

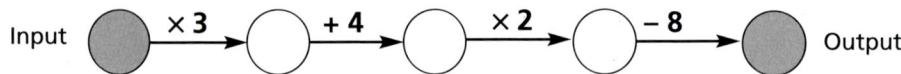

(b) Copy and complete this table for the arrow diagram above.

Input	Output
1	
3	
10	

(c) Look at your completed table.
What simple rule links the inputs and outputs?

(d) (i) Copy and complete the arrow diagram below for an input of n.
Simplify the output expression as far as you can.

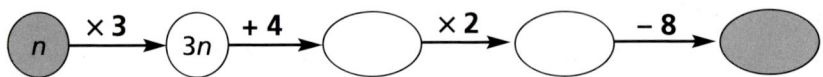

(ii) Explain how this shows the rule you found in (c) will work for **any** input.

B Multiplying out brackets 2

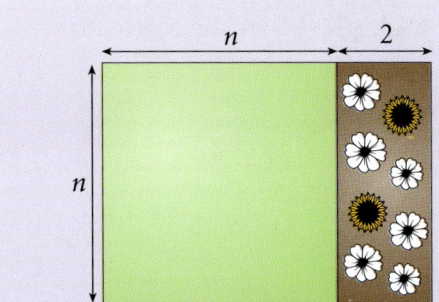
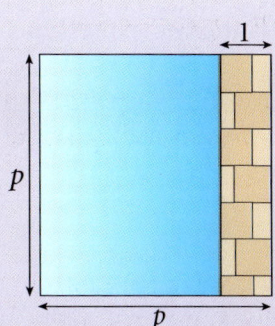

- What is the length of the whole garden?
- What is the area of the whole garden?

- What is the width of the pool?
- What is the area of the pool?

B1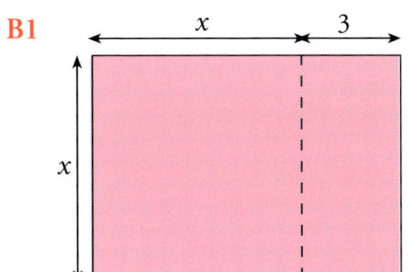

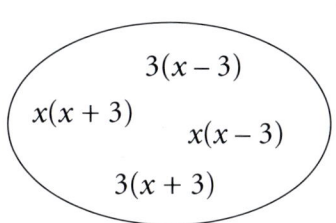

(a) Which expression in the loop gives the area of the pink rectangle?
(b) Write an expression without brackets for the area of the pink rectangle.
(c) Which expression in the loop gives the area of the yellow rectangle?
(d) Write an expression without brackets for the area of the yellow rectangle.

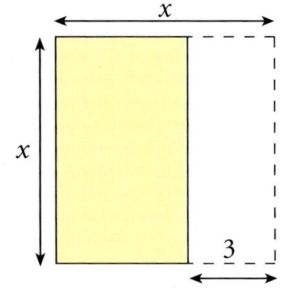

B2 Which of these expressions is equivalent to $n(n + 5)$?

$n^2 + 5$ $n^2 + 5n$ $2n + 5$

B3 Which of the following expressions is equivalent to $k(k - 7)$?

$k^2 - 7k$ $k^2 - 7$ $7k - k^2$ $7 - k^2$

B4 Multiply out the brackets in each of these.

(a) $n(n + 4)$
(b) $m(m + 7)$
(c) $x(x - 6)$
(d) $h(h - 4)$
(e) $b(3 + b)$
(f) $y(1 + y)$
(g) $a(5 - a)$
(h) $k(1 - k)$

B5 For each diagram
  (i) find the two expressions in the outer ring that **add** to give the centre expression
  (ii) find the two expressions in the outer ring that **multiply** to give the centre expression

(a) 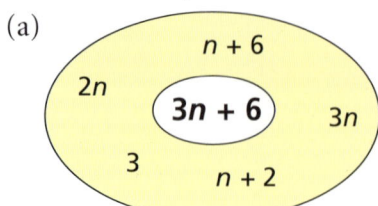 (b)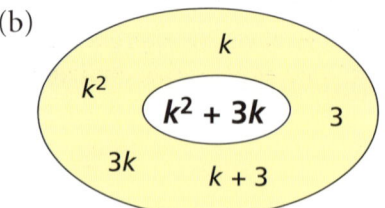

(c) 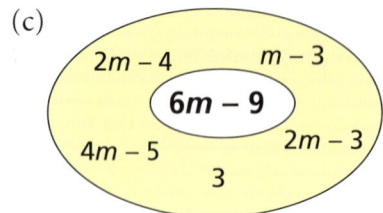 (d)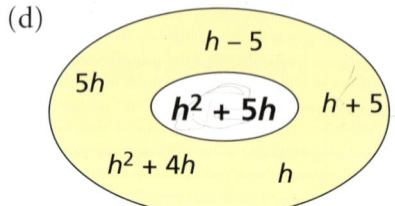

*B6 For each diagram
  • two expressions in the outer ring **add** to give the centre expression
  • two expressions in the outer ring **multiply** to give the centre expression
Can you find the expression that goes in each centre?

(a) 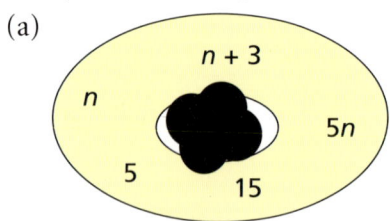 (b)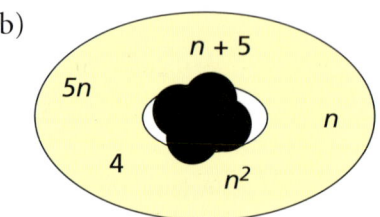

C Factorising

To factorise an expression, write it as a multiplication of simpler expressions.
Examples
  • $6a + 12 = 6(a + 2)$
  • $6a - 10 = 2(3a - 5)$
  • $a^2 + 5a = a(a + 5)$

We could write $6a + 12$ as $2(3a + 6)$ but we usually try to make the number outside the brackets as large as possible.

C1 Copy and complete each statement.
  (a) $3a + 6 = 3(a + \blacksquare)$ (b) $5a - 15 = 5(a - \blacksquare)$ (c) $4a + 20 = \blacksquare(a + 5)$

C2 Factorise these expressions.
(a) $2n + 14$ (b) $5n - 20$ (c) $3n + 21$ (d) $4n - 4$

C3 (a) Factorise $3n + 18$.
(b) Write down an expression for the length of one edge of this equilateral triangle.

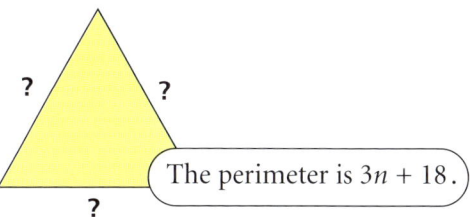

The perimeter is $3n + 18$.

C4 Copy and complete each statement.
(a) $6a + 9 = 3(2a + \blacksquare)$ (b) $10a - 5 = \blacksquare(2a - 1)$ (c) $8a + 20 = 4(\blacksquare + 5)$

C5 Factorise these expressions.
(a) $4n + 10$ (b) $9n - 6$ (c) $10n - 2$ (d) $15n + 25$

C6 (a) Factorise $8h + 12$.
(b) Write down an expression for the length of one edge of this square.

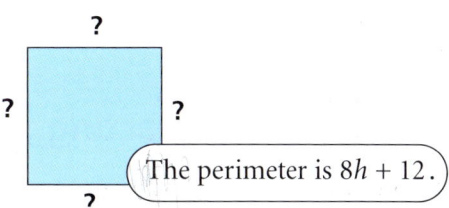

The perimeter is $8h + 12$.

C7 Copy and complete each statement.
(a) $a^2 + 6a = a(a + \blacksquare)$ (b) $k^2 - 3k = k(\blacksquare - 3)$ (c) $n^2 + 5n = n(\blacksquare + \blacksquare)$

C8 Factorise these expressions.
(a) $n^2 + 3n$ (b) $b^2 - 2b$ (c) $a^2 + a$ (d) $8x + x^2$

C9 Factorise these expressions.
(a) $12n + 8$ (b) $7n - 7$ (c) $n^2 + 8n$ (d) $n^2 - n$

D Fractions

n	$2n$	$3n$	$\frac{4n}{2}$	$\frac{6n}{2}$	$\frac{6n}{3}$	$\frac{8n}{4}$
1	2	3	2	3	2	2
2	4	6				
3						

D1 Which of these expressions are equivalent to $3x$?

$\dfrac{9x}{3}$ $\dfrac{12x}{3}$ $\dfrac{12x}{4}$ $\dfrac{15x}{3}$ $\dfrac{15x}{5}$

D2 Simplify these expressions.

(a) $\frac{10n}{2}$ (b) $\frac{12n}{6}$ (c) $\frac{20n}{4}$ (d) $\frac{10n}{5}$ (e) $\frac{18n}{3}$

D3 Copy and complete each statement.

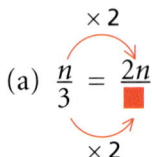

(a) $\frac{n}{3} = \frac{2n}{\blacksquare}$ (b) $\frac{n}{5} = \frac{\blacksquare}{15}$

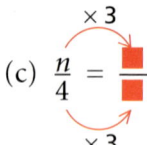

(c) $\frac{n}{4} = \frac{\blacksquare}{\blacksquare}$

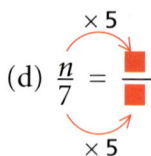

(d) $\frac{n}{7} = \frac{\blacksquare}{\blacksquare}$

D4 Which of these expressions are equivalent to $\frac{a}{4}$?

$\frac{2a}{8}$ $\frac{3a}{10}$ $\frac{4a}{16}$ $\frac{3a}{12}$ $\frac{5a}{10}$

D5 Copy and complete each statement.

(a) $\frac{n}{5} = \frac{2n}{\blacksquare}$ (b) $\frac{n}{3} = \frac{\blacksquare}{9}$ (c) $\frac{n}{2} = \frac{7n}{\blacksquare}$ (d) $\frac{n}{4} = \frac{\blacksquare}{20}$

E Adding and subtracting fractions

We can add or subtract fractions to give a single fraction.
Examples

$\frac{3}{5} + \frac{1}{5} = \frac{3+1}{5}$

$= \frac{4}{5}$

$\frac{1}{3} - \frac{x}{3} = \frac{1-x}{3}$

$\frac{1}{6} + \frac{x}{3} = \frac{1}{6} + \frac{2x}{6}$

$= \frac{1+2x}{6}$

E1 Simplify these by writing each as a single fraction.

(a) $\frac{1}{5} + \frac{2}{5}$ (b) $\frac{a}{5} + \frac{1}{5}$ (c) $\frac{3}{5} - \frac{b}{5}$ (d) $\frac{a}{5} + \frac{b}{5}$

E2 Copy and complete each statement.

(a) $\frac{a}{2} - \frac{1}{4} = \frac{\blacksquare}{4} - \frac{1}{4}$

$= \frac{\blacksquare}{4}$

(b) $\frac{2}{9} + \frac{x}{3} = \frac{2}{9} + \frac{\blacksquare}{9}$

$= \frac{\blacksquare}{9}$

E3 Which of these is equivalent to $\frac{1}{4} + \frac{k}{8}$?

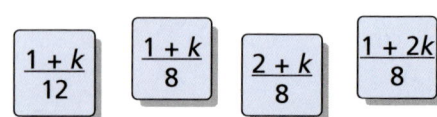

E4 Simplify each of these.

(a) $\frac{a}{2} + \frac{1}{8}$ (b) $\frac{1}{6} + \frac{b}{2}$ (c) $\frac{n}{10} + \frac{1}{5}$ (d) $\frac{a}{2} - \frac{b}{4}$

F True to form

In each statement, n can be any positive integer.

Which are
- always true?
- sometimes true and sometimes false?
- never true?

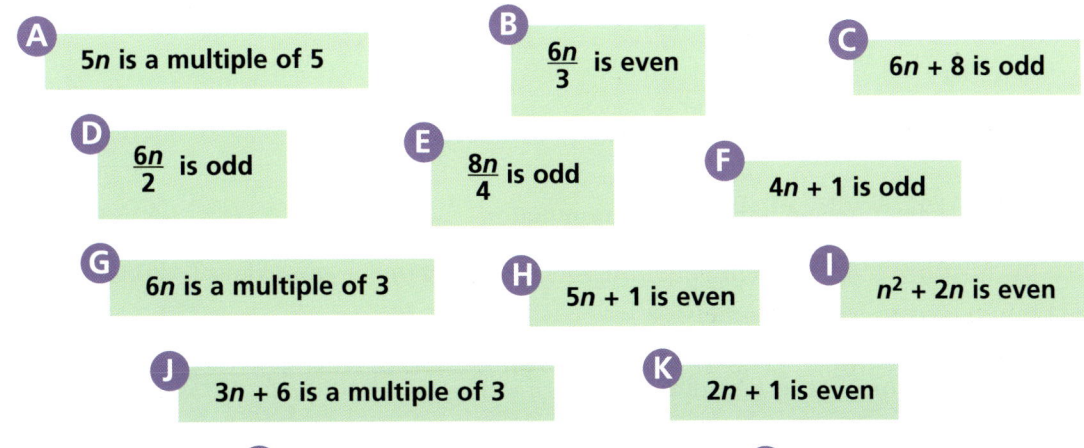

A $5n$ is a multiple of 5

B $\dfrac{6n}{3}$ is even

C $6n + 8$ is odd

D $\dfrac{6n}{2}$ is odd

E $\dfrac{8n}{4}$ is odd

F $4n + 1$ is odd

G $6n$ is a multiple of 3

H $5n + 1$ is even

I $n^2 + 2n$ is even

J $3n + 6$ is a multiple of 3

K $2n + 1$ is even

L $5n + 10$ is a prime number

M $n^2 + n$ is odd

What progress have you made?

Statement

I can multiply out brackets.

Evidence

1 Multiply out the brackets in each of these.
   (a) $2(n + 5)$ (b) $3(n - 2)$ (c) $4(5n + 3)$
   (d) $5(1 - 2n)$ (e) $n(n + 7)$ (f) $n(n - 3)$

I can factorise.

2 Factorise these expressions.
   (a) $2x + 14$ (b) $3x - 9$ (c) $6x + 4$
   (d) $10x - 15$ (e) $x^2 + 8x$ (f) $x^2 - 6x$

I can work with simple algebraic fractions.

3 Simplify these expressions.
   (a) $\dfrac{8n}{2}$ (b) $\dfrac{15p}{5}$ (c) $\dfrac{12t}{3}$

4 Simplify each of these.
   (a) $\dfrac{3}{4} + \dfrac{m}{4}$ (b) $\dfrac{n}{6} - \dfrac{1}{6}$
   (c) $\dfrac{1}{2} + \dfrac{a}{4}$ (d) $\dfrac{b}{5} - \dfrac{1}{10}$

3 Fractions

This work will help you
- add, subtract and multiply fractions
- use the inverse of fraction multiplication
- solve problems involving fractions

A Calculating with fractions – a reminder

It is easy to add fractions with the same denominator. $\frac{3}{20} + \frac{10}{20} = \frac{13}{20}$

When the fractions have different denominators, you need to use **equivalent fractions**.

Example $\frac{3}{8} + \frac{1}{6}$

For the denominator, choose a number which is a multiple of both 8 and 6. 24 is the smallest.

$$\frac{3}{8} = \frac{9}{24} \text{ and } \frac{1}{6} = \frac{4}{24}$$

So $\frac{3}{8} + \frac{1}{6} = \frac{9}{24} + \frac{4}{24} = \frac{13}{24}$.

A1 Work these out.
(a) $\frac{1}{5} + \frac{1}{3}$ (b) $\frac{1}{4} + \frac{1}{3}$ (c) $\frac{2}{5} + \frac{1}{3}$ (d) $\frac{3}{8} + \frac{2}{5}$ (e) $\frac{3}{10} + \frac{1}{6}$

A2 A spinner has three sections, red, blue and green.
The probability of getting red is $\frac{1}{4}$ and of getting blue is $\frac{1}{5}$.
Calculate the probability of getting (a) either red or blue (b) green

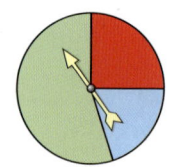

A3 A spinner has three sections, red, green and yellow.
The probability of getting red is $\frac{2}{3}$ and of green $\frac{1}{8}$.
Calculate the probability of getting (a) either red or green (b) yellow

A4 An old book on sign making gives this design for a letter L.
Work out its perimeter, giving your answer as a mixed number.
(The " symbol means inch or inches.)

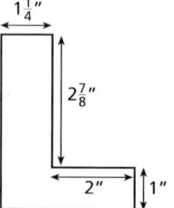

A5 Alan cuts a pizza into three pieces. He gives $\frac{1}{6}$ to Jade,
gives $\frac{1}{4}$ to Corin and keeps the rest himself.
(a) What fraction of the pizza does Alan get?
(b) What is the difference between Jade's fraction and Corin's?
(c) What is the difference between Alan's fraction and Jade's?

Egyptian fractions

The ancient Egyptians used only 'unit fractions' like $\frac{1}{2}$, $\frac{1}{3}$, $\frac{1}{4}$, $\frac{1}{5}$, ... (but they did allow $\frac{2}{3}$).
They expressed other fractions by adding unit fractions together.

For example, $\frac{5}{6} = \frac{1}{2} + \frac{1}{3}$ $\frac{19}{20} = \frac{1}{2} + \frac{1}{4} + \frac{1}{5}$

(They didn't repeat the same unit fraction, so $\frac{1}{3} + \frac{1}{3} + \frac{1}{5}$, for example, would not be allowed.)

- Find a way to write each of these fractions in the Egyptian way.
  (You may be able to find more than one way for some of them.)

  $\frac{5}{8}$ $\frac{7}{8}$ $\frac{7}{12}$ $\frac{17}{30}$ $\frac{4}{5}$

B Fraction of a fraction

You need a copy of sheet 254.

B1 Use square A on the sheet. The square represents 1 unit.

Outline $\frac{1}{2}$ of the square like this.   Then shade $\frac{1}{3}$ of the $\frac{1}{2}$ you outlined.

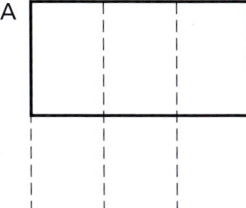 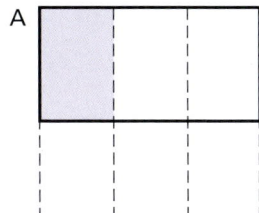

What fraction of the whole square have you shaded?

B2 Use square B on the sheet.
(a) Outline $\frac{1}{4}$ of the square.
(b) Shade $\frac{1}{5}$ of the $\frac{1}{4}$ you outlined.
(c) What fraction of the whole square have you shaded?

B3 Show these the same way.
For each one say what fraction of the whole square you have shaded.
(a) $\frac{1}{4}$ of $\frac{1}{2}$ on square C   (b) $\frac{1}{3}$ of $\frac{1}{8}$ on square D   (c) $\frac{1}{5}$ of $\frac{1}{5}$ on square E

B4 (a) Show $\frac{1}{3}$ of $\frac{1}{5}$ on square F.   (b) Show $\frac{1}{5}$ of $\frac{1}{3}$ on square G.
(c) What do you notice about $\frac{1}{3}$ of $\frac{1}{5}$ and $\frac{1}{5}$ of $\frac{1}{3}$?

B5 Show these on the unlettered squares and give their answers.
Use the 'ruler' cut from the edge of the sheet to mark the fractions.
(a) $\frac{1}{3}$ of $\frac{1}{6}$ (b) $\frac{1}{4}$ of $\frac{1}{4}$ (c) $\frac{1}{2}$ of $\frac{1}{5}$ (d) $\frac{1}{6}$ of $\frac{1}{4}$

B6 Give each missing fraction. Draw diagrams if you need to.
(a) $\frac{1}{2}$ of ? = $\frac{1}{16}$ (b) $\frac{1}{3}$ of ? = $\frac{1}{9}$ (c) ? of $\frac{1}{4}$ = $\frac{1}{12}$ (d) ? of $\frac{1}{2}$ = $\frac{1}{12}$

'of' and '×'

This shows 2 lots of $\frac{1}{3}$.

It makes sense to write $2 \times \frac{1}{3}$.

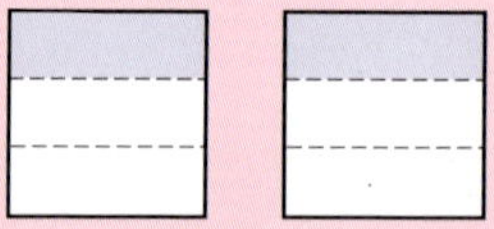

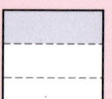

This shows $1\frac{1}{2}$ lots of $\frac{1}{3}$.

We can write $1\frac{1}{2} \times \frac{1}{3}$.

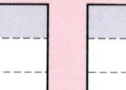

This shows 1 lot of $\frac{1}{3}$.

We can write $1 \times \frac{1}{3}$.

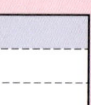

This shows $\frac{1}{2}$ a 'lot' of $\frac{1}{3}$.

It makes sense to write $\frac{1}{2} \times \frac{1}{3}$.

B7 Give answers to these. Draw on the unlettered squares if you need to.

(a) $\frac{1}{2} \times \frac{1}{2}$ (b) $\frac{1}{5} \times \frac{1}{6}$ (c) $\frac{1}{8} \times \frac{1}{4}$ (d) $\frac{1}{6} \times \frac{1}{6}$

B8 Write a multiplication (with its answer) for each of these.

(a) 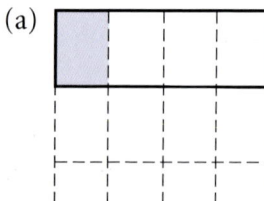   (b) 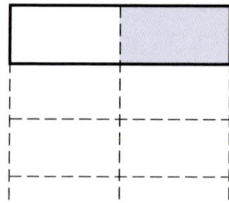   (c)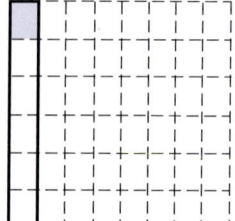

B9 A potter makes 80 coffee mugs. She colours half of them blue.

(a) How many does she colour blue?

She paints a hamster on a quarter of the blue ones.

(b) How many mugs are blue with a hamster on?

(c) What fraction of the 80 mugs are blue with a hamster on?

(d) How is this fraction related to the fractions given in the problem?

You need a fresh copy of sheet 254.

B10 Outline $\frac{3}{5}$ of square B.

Shade $\frac{1}{4}$ of the $\frac{3}{5}$ you outlined.

What fraction of the whole square have you shaded?

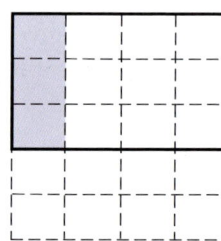

B11 Use square C on the sheet.
(a) Outline $\frac{1}{2}$ of the square.
(b) Shade $\frac{3}{4}$ of the $\frac{1}{2}$ you outlined.
(c) What fraction of the whole square have you shaded?

B12 Does $\frac{1}{3}$ of $\frac{5}{8}$ = $\frac{5}{8}$ of $\frac{1}{3}$?
Use squares D and H to explain.

B13 Show each of these by drawing and write its answer.
(a) $\frac{1}{5} \times \frac{3}{5}$ on square E (b) $\frac{2}{3} \times \frac{2}{5}$ on square F (c) $\frac{4}{5} \times \frac{2}{3}$ on square G

B14 Dave makes some waffles. He puts honey on $\frac{3}{5}$ of them.
Then he puts yogurt on $\frac{1}{4}$ of those that have honey on.
(a) What fraction of his waffles have honey on as well as yogurt?
(b) Given that he made 40 waffles, how many have honey on as well as yogurt?

B15 Describe how you can multiply one fraction by another fraction without drawing a diagram.

B16 Work these out, simplifying your answers where possible.
(a) $\frac{2}{3} \times \frac{2}{3}$ (b) $\frac{2}{3} \times \frac{3}{4}$ (c) $\frac{4}{5} \times \frac{1}{2}$ (d) $\frac{2}{3} \times \frac{5}{6}$

B17 Give each missing fraction.
(a) $? \times \frac{5}{6} = \frac{25}{36}$ (b) $\frac{3}{4} \times ? = \frac{9}{32}$ (c) $? \times \frac{7}{10} = \frac{21}{40}$ (d) $\frac{3}{5} \times ? = \frac{21}{40}$

You know that simplifying an answer to a fraction multiplication involves dividing top and bottom by the same number.

$\frac{1}{4} \times \frac{2}{9} = \frac{2}{36} \xrightarrow[\div 2]{\div 2} \frac{1}{18}$

Sometimes you can spot one number to divide top and bottom by before you multiply the fractions. This is called **cancelling**.

$\frac{1}{\cancel{4}_2} \times \frac{\cancel{2}^1}{9} = \frac{1}{18}$

Sometimes cancelling involves two or more stages. $\frac{3}{8} \times \frac{\cancel{4}^1}{9}_2 \Rightarrow \frac{\cancel{3}^1}{\cancel{8}_2} \times \frac{\cancel{4}^1}{\cancel{9}_3} = \frac{1}{6}$

B18 Give answers to these, cancelling first where possible.
(a) $\frac{5}{6} \times \frac{3}{4}$ (b) $\frac{1}{6} \times \frac{3}{8}$ (c) $\frac{5}{6} \times \frac{2}{5}$ (d) $\frac{2}{3} \times \frac{3}{10}$

***B19** Give each missing fraction.
(a) $? \times \frac{4}{5} = \frac{3}{5}$ (b) $\frac{3}{5} \times ? = \frac{2}{5}$ (c) $? \times \frac{5}{8} = \frac{1}{2}$ (d) $\frac{9}{10} \times ? = \frac{3}{4}$

C Flows

C1 Plain buns come into a cake factory.
$\frac{1}{2}$ of them are filled with cream and $\frac{1}{2}$ of them are filled with jam.

$\frac{1}{2}$ of the cream buns get pink icing and $\frac{1}{2}$ get white icing.
$\frac{3}{4}$ of the jam buns get pink icing and $\frac{1}{4}$ get white icing.

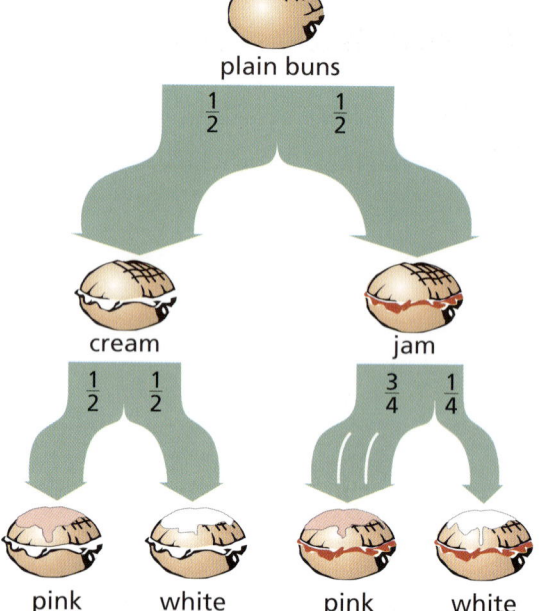

(a) What fraction of the plain buns become
   (i) cream buns with pink icing   (ii) cream buns with white icing
   (iii) jam buns with pink icing   (iv) jam buns with white icing
(b) What do the four fractions in your answer to (a) add up to?
(c) Every hour, 240 plain buns come into the factory. How many become
   (i) cream buns with pink icing   (ii) cream buns with white icing
   (iii) jam buns with pink icing   (iv) jam buns with white icing
   Check that your answers add up to 240.

C2 This diagram shows how the traffic leaving a town splits into different routes.

When the traffic gets to A it splits up. $\frac{1}{3}$ goes to B and $\frac{2}{3}$ goes to C.

The traffic arriving at B splits $\frac{4}{5}$ to D and $\frac{1}{5}$ to E.

So the traffic arriving at D is $\frac{4}{5}$ of $\frac{1}{3}$ of the total, which is $\frac{4}{15}$ of the total.

What fraction of the total traffic arrives at
(a) E   (b) F   (c) G

Check that the fractions arriving at D, E, F and G add up to 1.

D Thinking backwards

D1 You can see $\frac{3}{4}$ of this window. How many panes are there in the whole window?

D2 You can see $\frac{2}{5}$ of this train.

How many carriages are there in the whole train?

D3 In a bag of beads, 15 of the beads are blue.
The probability of getting a blue when selecting a bead at random is $\frac{3}{5}$.
How many beads are there in the bag altogether?

D4 In a bag of beads, $\frac{1}{4}$ are red, $\frac{3}{8}$ are yellow, $\frac{1}{6}$ are blue and the rest are green.
  (a) What fraction of the beads are green?
  (b) Given that there are 10 green beads, how many beads are there altogether in the bag?

D5 Gail is weaving rugs, which are all identical.
In $\frac{1}{4}$ hour she weaves $\frac{2}{3}$ of a rug.
How many complete rugs will she weave in 4 hours, if she works at the same rate?

D6 A gardening club held an election to decide this year's president.
$\frac{2}{3}$ of the members voted.
Paula was elected president with $\frac{5}{8}$ of the votes.
She got 40 votes. How many members did the club have?

D7 Hywel bought a large bar of chocolate.
He gave away $\frac{3}{4}$ of the bar and ate $\frac{5}{7}$ of the rest.
This is what he had left.

How many squares were there in the whole bar of chocolate?

*D8 This diagram shows what happens to the traffic leaving a city.
Traffic splits at A, B and C.

The fractions of the total traffic reaching
D, E, F and G are shown.

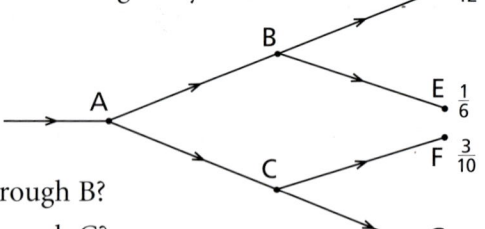

(a) What fraction of the total traffic goes through B?
(b) What fraction of the total traffic goes through C?
(c) What fraction of the traffic reaching B goes to D?
(d) What fraction of the traffic reaching C goes to F?

What progress have you made?

Statement

Evidence

I can add, subtract and multiply fractions.

1 Work these out.
   (a) $\frac{5}{8} - \frac{1}{4}$ (b) $\frac{1}{5} + \frac{2}{15}$ (c) $\frac{5}{6} - \frac{4}{5}$

2 Work these out, cancelling where possible.
   (a) $\frac{1}{3} \times \frac{1}{10}$ (b) $\frac{2}{5} \times \frac{3}{8}$ (c) $\frac{3}{10} \times \frac{5}{6}$

I can solve problems and think backwards with fractions.

3 Jacob gives away $\frac{5}{9}$ of a cake and eats $\frac{1}{4}$ of what he keeps.
   What fraction of the whole cake is left?

4 Give the missing fractions.
   (a) $\frac{2}{3} \times ? = \frac{4}{9}$ (b) $? \times \frac{3}{10} = \frac{9}{40}$ (c) $? \times \frac{3}{8} = \frac{1}{8}$

5 Tiffany owns $\frac{1}{3}$ of a field.
   She grows mustard in $\frac{4}{5}$ of her part and kale in the rest.
   Melanie owns the other $\frac{2}{3}$ of the field.
   She grows broccoli in $\frac{3}{4}$ of her part and cauliflowers in the rest.
   What fraction of the field is
   (a) mustard   (b) kale
   (c) broccoli   (d) cauliflowers

6 A lorry driver has driven 162 miles this morning.
   This is $\frac{3}{8}$ of his complete journey.
   How long is his complete journey?

4 As time goes by

The work will help you
- draw and interpret graphs that represent real-life situations

A Filling up

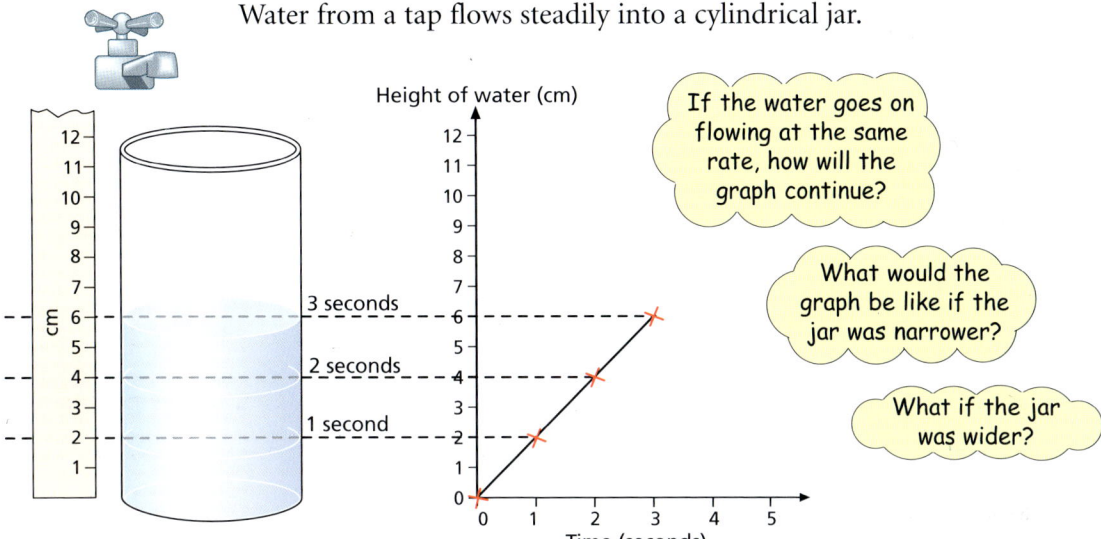

Water from a tap flows steadily into a cylindrical jar.

If the water goes on flowing at the same rate, how will the graph continue?

What would the graph be like if the jar was narrower?

What if the jar was wider?

A1 These diagrams show the level of water in a jar at various times.

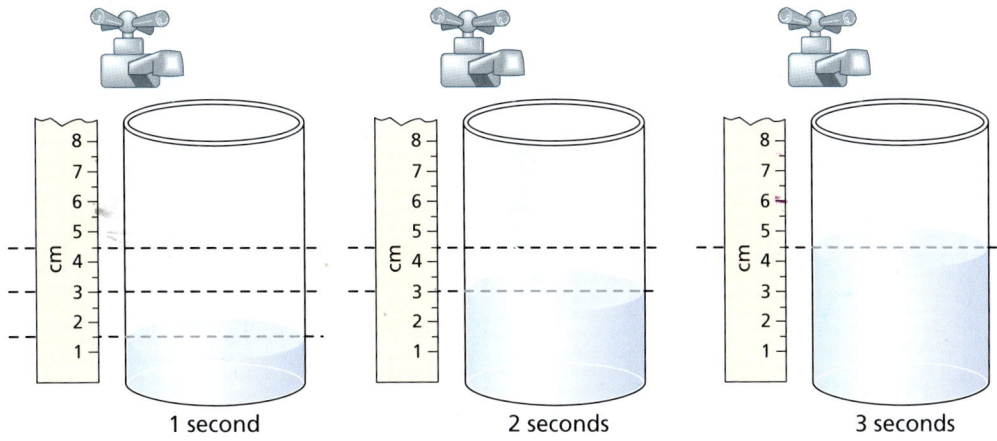

(a) What is the level of water after 3 seconds?
(b) If water continues to flow steadily, what will the water level be after 5 seconds?
(c) On graph paper, draw a graph to show the water level during these 5 seconds.
(d) Use your graph to estimate when the level of water in the jar reaches 5.5 cm.

A2 For six seconds, water flows steadily at the same rate into two cylindrical jars. The graph shows the water level in each jar during this time.

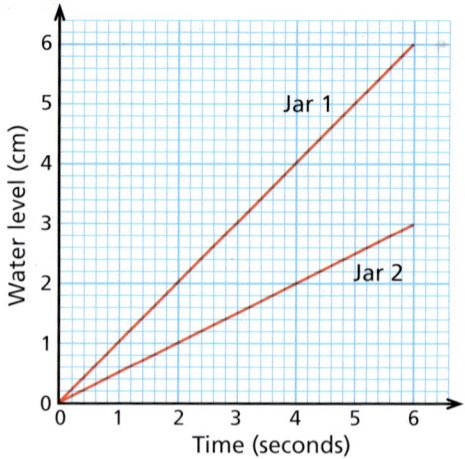

(a) What is the level of water in jar 1 after 3 seconds?
(b) What is the level of water in jar 2 after 6 seconds?
(c) What is the level of water in each jar after 4 seconds?
(d) Which jar is narrower, jar 1 or jar 2?
Explain your answer.

A3 Water flows steadily at the same rate into these cylindrical jars. Match each jar to one of the graphs.

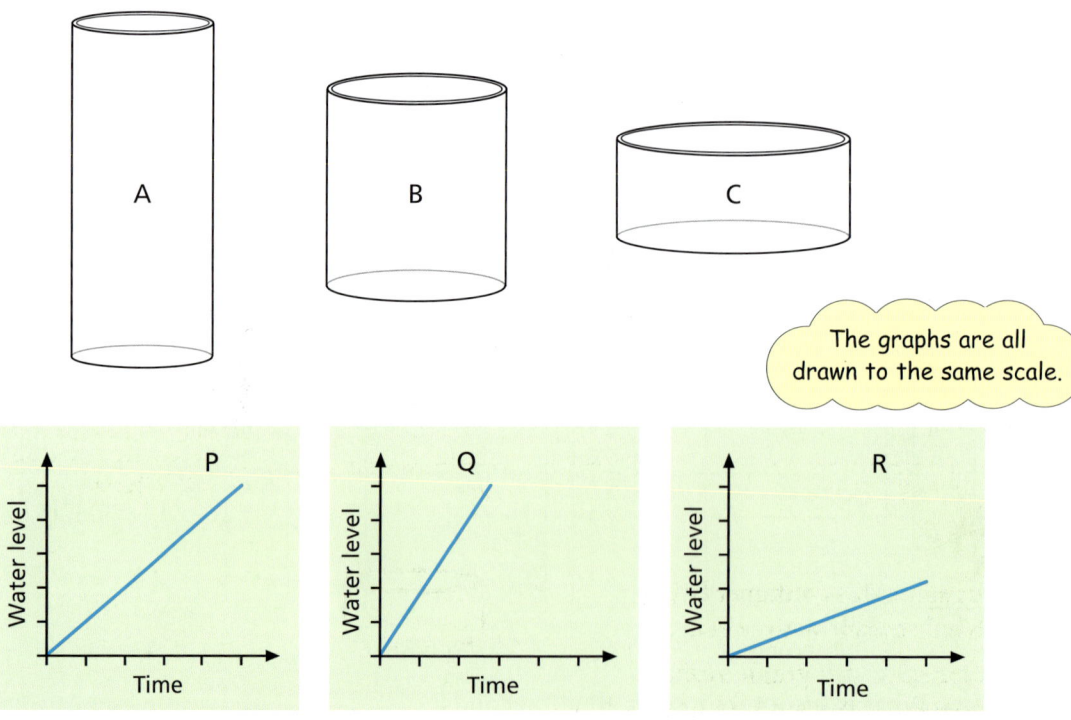

The graphs are all drawn to the same scale.

A4 Imagine this container being filled steadily with water.

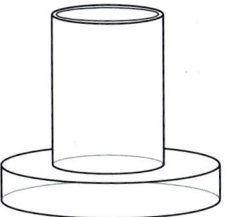

Which graph shows the water level in the container as it is filled?

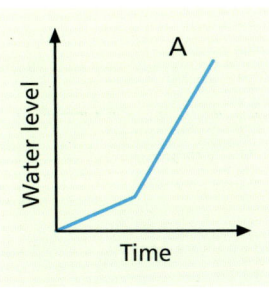 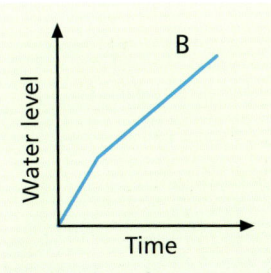 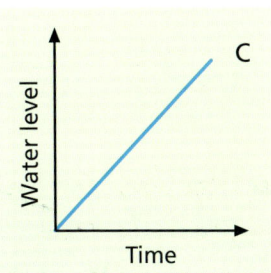

A5 Water flows steadily at the same rate into these containers. Match each container to one of the graphs below.

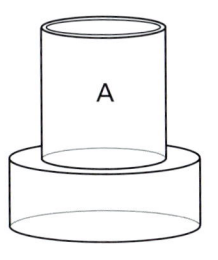 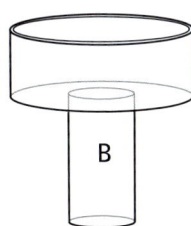 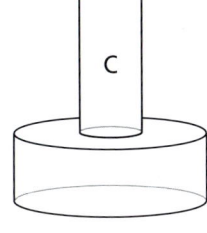

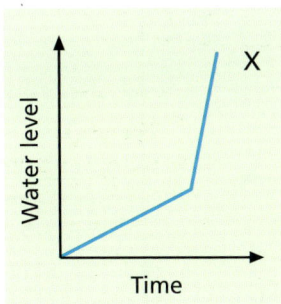 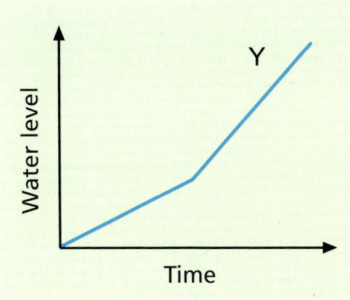 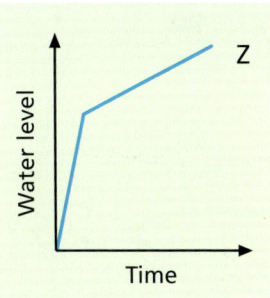

A6 Imagine this container being filled steadily with water.

Draw a sketch graph to show how the water level changes as it fills.

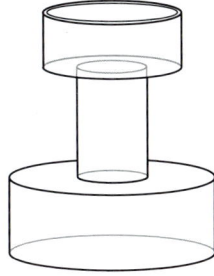

B More filling

Water from a tap flows steadily into a conical container.

What will the graph of the water level look like?

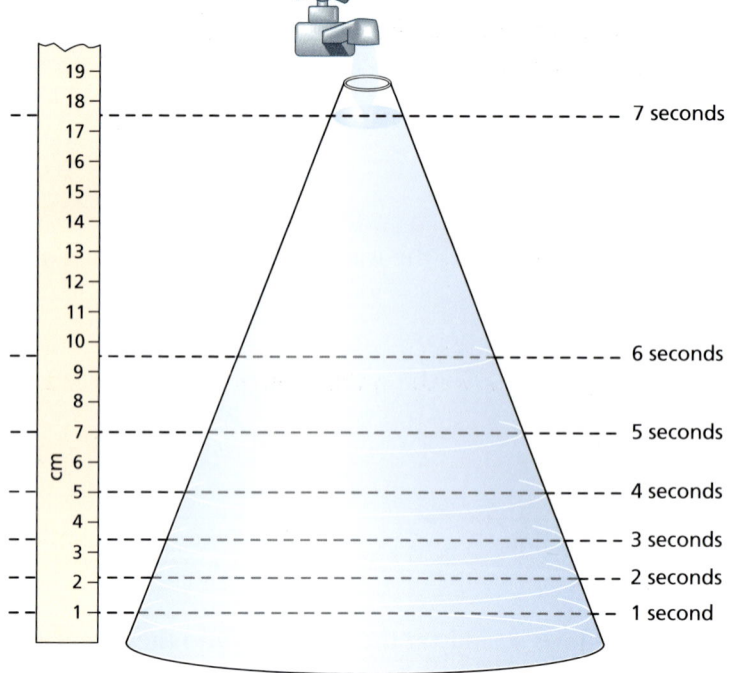

B1 Imagine each of these containers being filled steadily at the same rate.
Match each container to its graph.

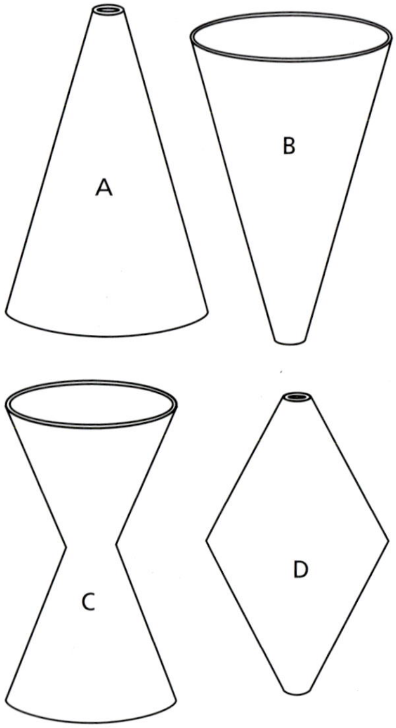

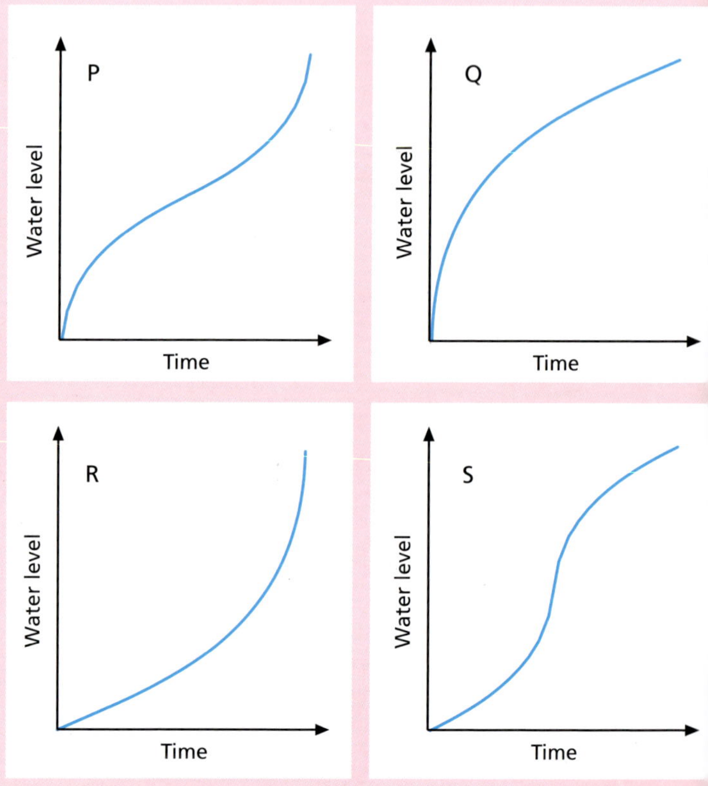

30

B2 Imagine this container being filled steadily with water.

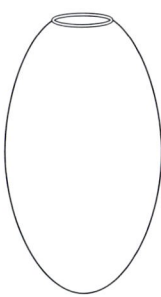

Which graph shows the water level in the container as it is filled?

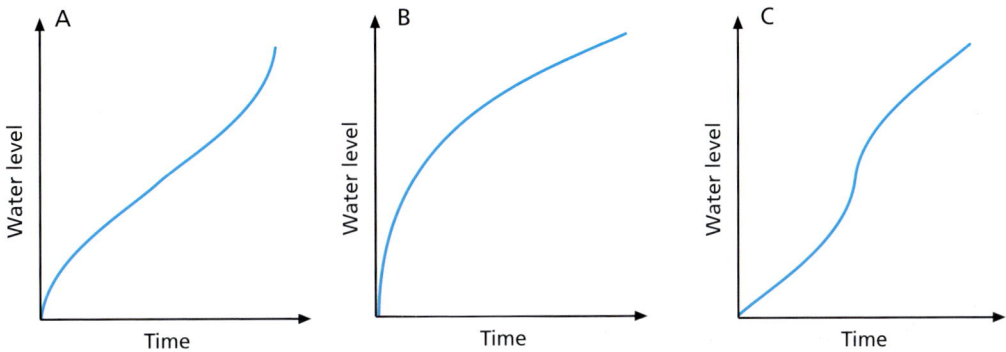

B3 Imagine each of these containers being filled from a steady tap. Draw sketch graphs to show how the water level changes.

(a) 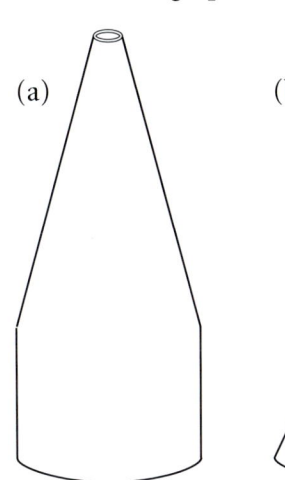   (b) 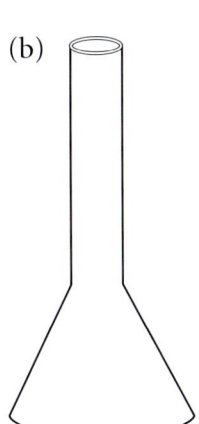   (c)    (d)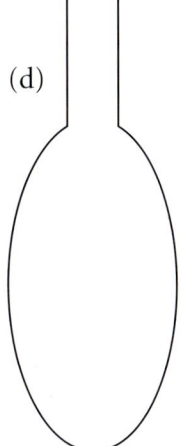

C Diamonds are forever?

Police believe that Sad Harry broke into the safe at De Vere's in Stevenage last night. It is alleged that he stole all their diamonds and then set off at exactly midnight from Stevenage up the Great North Road at 40 miles per hour so as not to attract attention.

C1 How far does Harry travel in the first hour of his journey?

C2 (a) Stevenage is 50 miles north of London.

  (i) How far would Harry be from London at 1 a.m.?

  (ii) Copy and complete the table below for Harry's journey.

Time	Distance from London (miles)
Midnight	50
1 a.m.	
2 a.m.	
3 a.m.	

(b) Draw a graph for Harry's journey. Use axes like these. (Go as far as 3 a.m.)

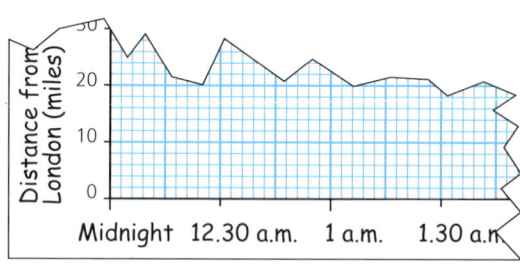

C3 The manageress of De Vere's, Minnie Sparkler, was in London at the time. She leapt into her car and set off at midnight at a steady speed up the Great North Road. By 2 a.m. she had travelled 120 miles. In fact, she helped Harry plan the whole thing and wanted to get to the hideout before him.

(a) Draw a graph for Minnie's journey from London on the same axes as Sad Harry's. (Go as far as 3 a.m.)

(b) What speed was Minnie travelling at?

C4 (a) How far was Harry from London after half an hour?

(b) How far from London was Minnie at the same time?

(c) How far apart were the two cars?

(d) At what time did Minnie pass Sad Harry?

C5 The police did not hear of the raid until 1.30 a.m. A car was immediately sent from London, and travelled north at a speed of 120 m.p.h.

The police wanted to speak to Minnie and to Sad Harry.
Whose car did they catch up with first?
(You will need to draw another line on your graph.)

D Speed graphs

Jude is a keen cyclist.
She always keeps an eye on the speed she is going at.

When she cycles to school, she goes along Clark Road, Beacon Road, Hill View Road and School Road.

This graph shows her speed on the way to school.

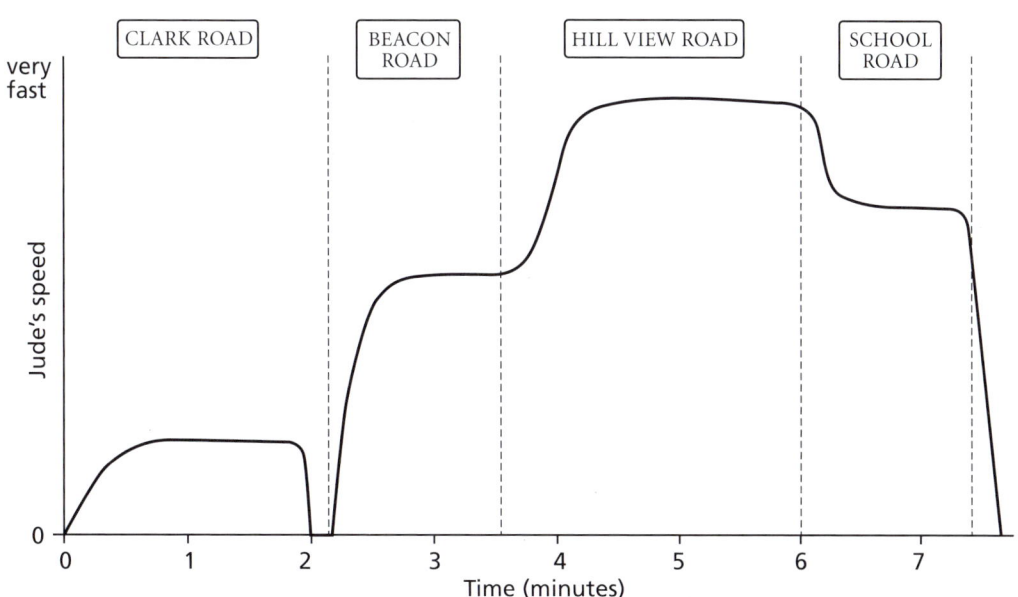

You can see that Jude cycled quite slowly along Clark Road,
But she went more quickly along most of Beacon Road.

D1 (a) Along which road did Jude cycle fastest?

(b) Along which road did she cycle second fastest?

(c) There are traffic lights at the end of one road.
Jude had to stop at the lights.

(i) Which road do you think the traffic lights are at the end of?

(ii) How many minutes does it take Jude to cycle from home to the traffic lights?

D2 Masud went for a cycle ride one day.
This is a graph of his speed on the ride.

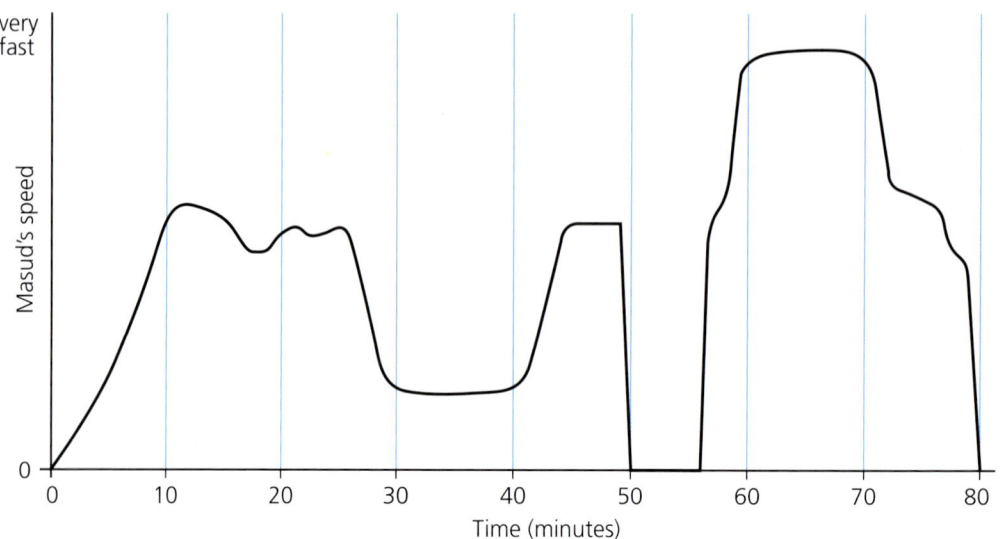

(a) At one time on his ride Masud stopped to talk to a friend.
How many minutes had he been cycling for when he stopped?

(b) For 10 minutes Masud was cycling slowly up a hill.
Between what times was he cycling uphill?

(c) For 10 minutes, Masud was cycling very fast downhill.
Between what times was he cycling downhill?

D3 Callie is a sprinter.
Her trainer has written down instructions for a training run.

> **Training: Callie**
>
> First jog for about 10 minutes.
> Then run as fast as you can for 5 minutes,
> followed by jogging for 5 minutes.
> Then walk for 5 minutes.
>
> Rest sitting for 5 minutes, then jog for 5 minutes
> and then sprint for 5 minutes.
> Walk for 5 minutes and then you can stop.
>
> See you Wednesday,
> Fred

Sketch a graph of Callie's speed as she trains.

Label your up (speed) axis **0** at the bottom and **very fast** at the top.
Your across (time) axis will need to go from 0 to 50 minutes.

D4 Derek goes for a bike ride.

First he cycles at a steady rate.
Then he goes steadily but slowly up a long hill.
Lastly he goes quickly down a steep hill, and then stops.

(a) Which of the sketch graphs do you think shows his speed?

(b) Make up your own stories for the other two graphs.

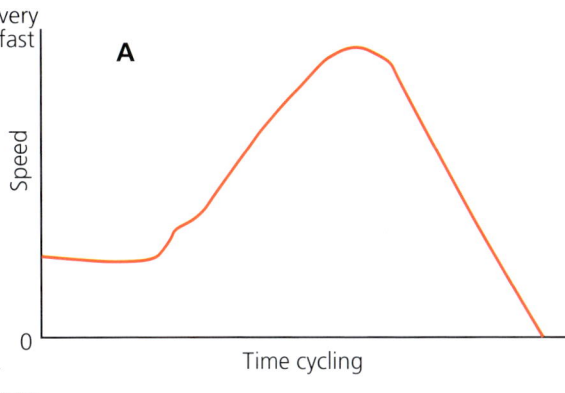

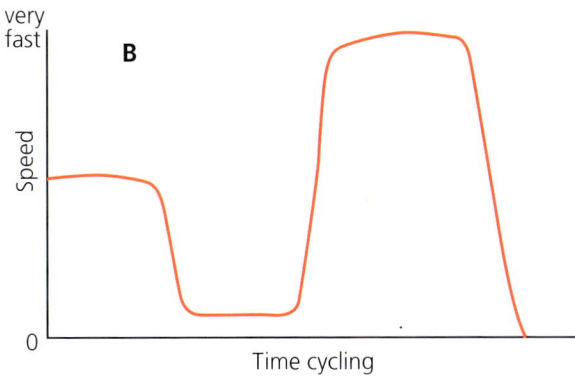

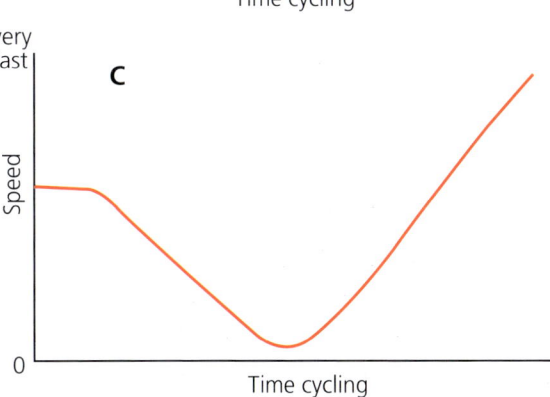

D5 Della is trying out a friend's moped to see if she wants to buy it.

She takes it for a test run and tells her friend about it.

Well, it went OK for about 5 minutes, and then I got held up in the High Street for 5 minutes.

It took me ages to get up Quarry Bank – about 10 minutes, the hill was so steep. Coming down the other side it went like the wind for about 5 minutes, but then it just conked out completely.

I let it cool off for 10 minutes, but then I couldn't start it and it took me 10 minutes to push it back here!

Draw a sketch graph of her speed on the moped.

What progress have you made?

Statement

I can interpret real-life graphs.

Evidence

1 (a) Water flows steadily into these containers. Match each container to a graph.

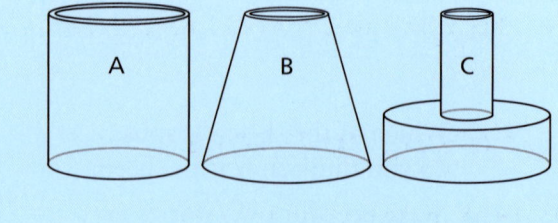

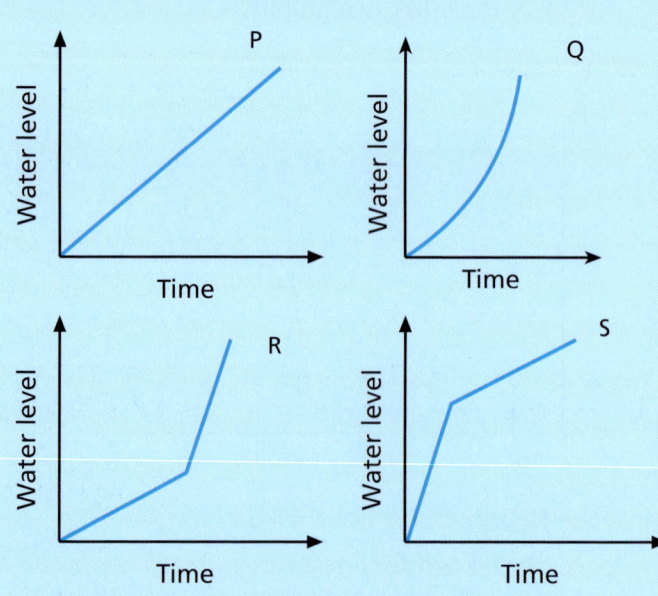

(b) Sketch a suitable container for the unmatched graph.

I can sketch real-life graphs.

2 Sketch a speed graph for this story. Label your speed axis **0** at the bottom and **very fast** at the top. Your time axis will go from 0 to 60 minutes.

> Went out cycling – I was out for an hour altogether. First I cycled up Crudge Bank – took me 10 minutes, going really slowly. Then I stopped for 10 minutes! But then top speed down the other side for 5 whole minutes! Then I met Jim walking and pushed my bike alongside him for 10 minutes.
>
> After that, I cycled steadily for 15 minutes, and then sprinted for 10 minutes all the way home.

5 Working with rules

This work will help you
- solve simple equations
- substitute numbers into a formula to give an equation and solve it
- change the subject of a simple formula

A Solving equations

You can solve the equation $\frac{n}{4} + 1 = 6$ ...

... by balancing ...

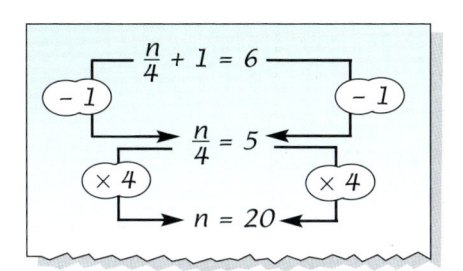

$\frac{n}{4} + 1 = 6$
$-1 \qquad -1$
$\frac{n}{4} = 5$
$\times 4 \qquad \times 4$
$n = 20$

... or by drawing a flow diagram and reversing it.

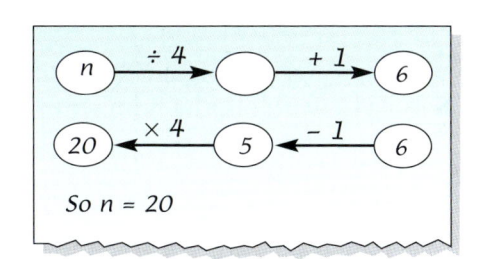

$n \xrightarrow{\div 4} \bigcirc \xrightarrow{+1} 6$

$20 \xleftarrow{\times 4} 5 \xleftarrow{-1} 6$

So $n = 20$

A1 Solve these equations.
(a) $2n + 1 = 15$
(b) $3p + 5 = 32$
(c) $5 + 2m = 12$
(d) $14 = 5k - 2$

A2 Solve these equations.
(a) $\frac{n}{2} + 1 = 4$
(b) $\frac{m}{3} + 7 = 10$
(c) $\frac{p}{4} + 3 = 9$
(d) $2 + \frac{x}{3} = 4$
(e) $12 = \frac{x}{5} + 2$
(f) $\frac{y}{3} - 3 = 5$
(g) $\frac{z}{6} - 2 = 3$
(h) $\frac{w}{9} - 4 = 0$

A3 Solve these equations.
(a) $5x + 6 = 13$
(b) $\frac{y}{5} + 6 = 8$
(c) $3z - 8 = 10$
(d) $\frac{p}{7} - 1 = 2$
(e) $15 = 9 + 4m$
(f) $15 = \frac{n}{9} + 10$
(g) $37 = 8k - 3$
(h) $\frac{h}{4} - 5 = 3$

B From formula to equation

Superstore Systems make racks to fit any number of compact discs.
The formula for the height of a rack is

$H = 14d + 60$

where H is the height of the rack in mm and d is the number of discs.

- What is the height of a rack that holds 10 discs?
- The height of another rack is 410 mm.
  Form an equation and solve it to find the number of discs this rack holds.
- Find the value of d when $H = 830$.

B1 Superstore Systems make another style of rack.
The formula for the height of one of these racks is

$H = 7d + 40$

where H is the height of the rack in mm and d is the number of discs.

(a) What is the height of a rack that holds 20 discs?

(b) The height of another of these racks is 320 mm.
Form an equation and solve it to find the number of discs this rack holds.

(c) Find the value of d when $H = 390$.

B2 The Rugged Walk outdoor centre organises walking trips.
They use this formula to work out the number of maps to take for groups of walkers.

$m = \frac{w}{2} + 1$

where m is the number of maps and w is the number of walkers in the group.

(a) A group of walkers takes 10 maps.
Form an equation and solve it to find the number of walkers in the group.

(b) Find the value of w when $m = 15$.

B3 The rule for this pattern is

$m = 6n - 2$

where m is the number of matches
and n is the pattern number.

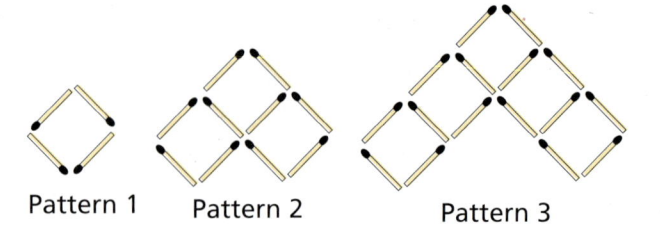

Pattern 1   Pattern 2   Pattern 3

(a) Use the rule to work out the number of matches in pattern 4.

(b) One of these patterns uses 112 matches.
Form an equation and solve it to find which pattern this is.

(c) Find the value of n when $m = 166$.

C Inverse rules

The rule for this pattern is
$$m = 6n - 2$$
where m is the number of matches and n is the pattern number.

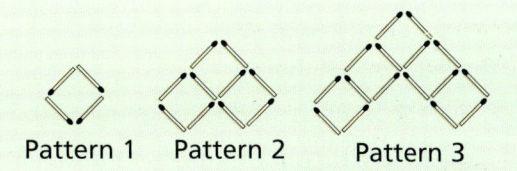

Pattern 1   Pattern 2   Pattern 3

The flow diagram for this rule is

$n \xrightarrow{\times 6} \bigcirc \xrightarrow{-2} m$

Reversing the flow diagram gives this rule for n in terms of m.

$\boxed{\frac{m+2}{6}} \xleftarrow{\div 6} (m+2) \xleftarrow{+2} m$

$$n = \frac{m+2}{6}$$

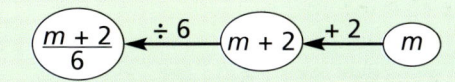

C1 The Rugged Walk outdoor centre use this formula to work out the number of sandwiches to take for groups of walkers.

$$S = 3w + 5$$

where S is the number of sandwiches and w is the number of walkers in the group.

(a) Decide which of these is the flow diagram for this formula and copy it.

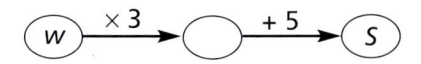

$w \xrightarrow{\times 3} \bigcirc \xrightarrow{+5} S$ $w \xrightarrow{+5} \bigcirc \xrightarrow{\times 3} S$

(b) (i) Reverse your flow diagram and find a formula for w in terms of S. Your formula should begin $w = \ldots$.

(ii) Use your formula to work out the number of walkers in a group that takes 53 sandwiches.

C2 (a) Draw a flow diagram for the formula
$$P = 5N - 3$$
Your diagram should begin $N \longrightarrow$

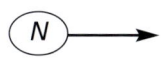

(b) Reverse the flow diagram and find a formula for N in terms of P. Your formula should begin $N = \ldots$.

(c) Work out the value of N when
  (i) $P = 27$ (ii) $P = 97$ (iii) $P = 10$

C3 Rearrange these to find a formula for N in terms of P each time. Each formula should begin $N = \ldots$.
(a) $P = 2N + 1$ (b) $P = 4N + 3$ (c) $P = 5N - 1$
(d) $P = 7N - 3$ (e) $P = 2 + 3N$ (f) $P = 10N - 9$

C4 The rule for this pattern is
$$d = 2n - 1$$
where d is the number of dots and n is the pattern number.

Pattern 1   Pattern 2   Pattern 3

(a) Work out the number of dots in pattern 10.

(b) (i) Find a formula for n in terms of d.

(ii) Use this new formula to find which of these patterns uses 133 dots.

C5 Sometimes Busby's restaurant arrange their standard rectangular tables in a line like this.

(a) If they use 10 tables in a line, how many chairs do they need?

(b) Work out how many chairs they would need for 100 tables in a line!

(c) Which of these formulas tells you the number of chairs (c) when you know the number of tables (t)?

$c = t + 2$ $c = 4t + 2$ $c = 6t$ $c = t + 4$ $c = 2t + 4$

(d) (i) Find a formula for t in terms of c.

(ii) Use your formula to find the number of tables you would need to seat 58 people.

C6 Here is another arrangement at Busby's using larger tables.

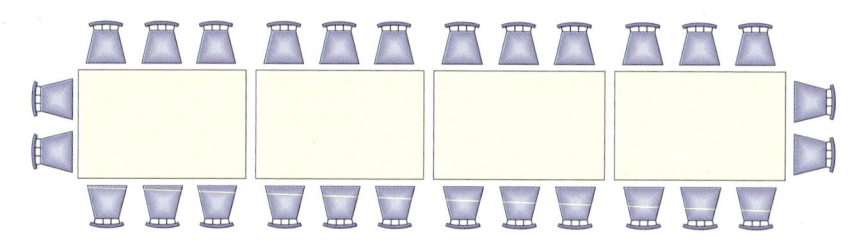

(a) How many chairs does this arrangement need for 20 tables in a line?

(b) Write a formula to tell you the number of chairs (c) when you know the number of tables (t).

(c) (i) Rearrange your formula to give a formula for t in terms of c.

(ii) Use your formula to find the number of tables you would need to seat 58 people.

C7 A formula that links x and y is $y = \frac{x+3}{2}$.

(a) Decide which of these is its flow diagram and copy it.

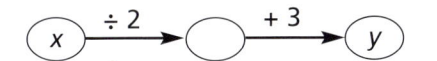

$x \xrightarrow{\div 2} \bigcirc \xrightarrow{+3} y$ $\qquad$ $x \xrightarrow{+3} \bigcirc \xrightarrow{\div 2} y$

(b) (i) Reverse the flow diagram and find a formula for x in terms of y.
(ii) Use your formula to work out the value of x when $y = 20$.

C8 Rearrange these to find a formula for p in terms of q each time.

(a) $q = \frac{p+1}{3}$ $\qquad$ (b) $q = \frac{p-3}{2}$ $\qquad$ (c) $q = \frac{p-1}{5}$

D Using brackets

The flow diagram for the rule

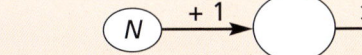

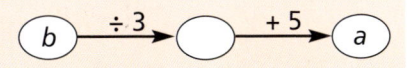

$P = 2(N+1)$ is $\quad N \xrightarrow{+1} \bigcirc \xrightarrow{\times 2} P$

Reversing the flow diagram gives $\quad \frac{P}{2} - 1 \xleftarrow{-1} \frac{P}{2} \xleftarrow{\div 2} P$

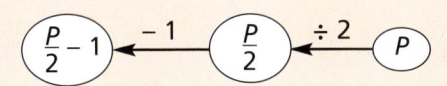

$N = \frac{P}{2} - 1$

The flow diagram for the rule

$a = \frac{b}{3} + 5$ is $\quad b \xrightarrow{\div 3} \bigcirc \xrightarrow{+5} a$

Reversing the flow diagram gives $\quad 3(a-5) \xleftarrow{\times 3} a-5 \xleftarrow{-5} a$

$b = 3(a-5)$

D1 Rearrange these to find a formula for N in terms of S each time.
(a) $S = 3(N+4)$ $\qquad$ (b) $S = 5(N-1)$ $\qquad$ (c) $S = 2(N-6)$

D2 Rearrange these to find a formula for x in terms of y each time.
(a) $y = \frac{x}{3} - 5$ $\qquad$ (b) $y = \frac{x}{2} + 1$ $\qquad$ (c) $y = \frac{x}{10} - 3$

D3 Rearrange these to find a formula for t in terms of v each time.
(a) $v = 4(t+5)$ $\qquad$ (b) $v = \frac{t}{4} + 5$ $\qquad$ (c) $v = 3(t-9)$

(d) $v = \frac{t}{5} - 5$ $\qquad$ (e) $v = \frac{t}{7} + 6$ $\qquad$ (f) $v = 2(t-10)$

D4 A shop uses this rule to work out the length of fabric needed for some curtains.

> Add 15 to the height of the window in centimetres and then multiply the result by 4.

(a) Use the rule to work out the length of fabric needed for a window of height 85 cm.

(b) Which of these formulas links the length of fabric (L) and the height of the window (H)?

$$L = 4H + 15 \qquad L = 15H + 4 \qquad L = 4(H + 15)$$

(c) (i) Find a formula for H in terms of L.

(ii) The shop has a piece of curtain fabric that is 460 cm in length. Joy uses all this fabric to make curtains for her bedroom window.
What is the height of the window?

D5 The Rugged Walk outdoor centre uses this rule to work out the number of survival bags to take for groups of walkers.

> Take 1 bag for each pair of walkers and 1 extra bag.

(a) Use the rule to work out the number of survival bags they take for
   (i) 6 walkers (ii) 10 walkers (iii) 20 walkers

(b) Find a formula for the number of bags (b) in terms of the number of walkers (w). Your formula should begin $b = \ldots$.

(c) (i) Find a formula for w in terms of b.

(ii) The centre has a total of 20 survival bags.
What is the largest possible group of walkers the centre can take out?

E Formulas with several letters

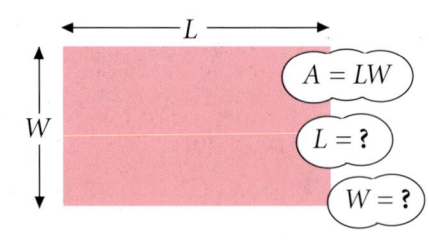

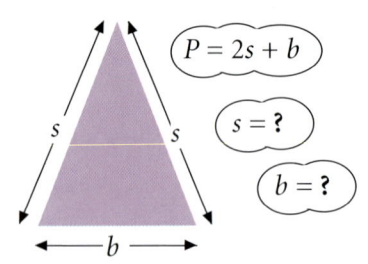

$A = LW$
$L = ?$
$W = ?$

$P = 2s + b$
$s = ?$
$b = ?$

E1 Rearrange each formula to make N the subject.
Each formula should begin $N = \ldots$.

(a) $P = N + M$
(b) $Q = N - A$
(c) $A = MN$
(d) $B = 3N + P$
(e) $H = 6N - K$
(f) $Y = \dfrac{N}{3} + M$
(g) $C = \dfrac{N}{5} - T$
(h) $R = \dfrac{N + Y}{4}$
(i) $F = \dfrac{N - T}{8}$

E2 The perimeter of a rectangle is given by the formula
$$P = 2(L + W)$$
where L is the length and W is the width.

(a) Write the formula with L as the subject.

(b) Use your formula to find the length of a rectangle that has a width of 6.3 cm and a perimeter of 28.2 cm.

E3 Rearrange each formula so that the letter shown in red is the subject.

(a) $y = 3(\mathbf{a} + b)$
(b) $k = 4(\mathbf{x} - y)$
(c) $p = m\mathbf{t} - 5$
(d) $v = \mathbf{a}t + u$
(e) $y = \dfrac{\mathbf{p} + q}{6}$
(f) $x = \dfrac{h}{\mathbf{n}} - p$

E4 Rearrange each formula to make N the subject.

(a) $Y = N + 2M$
(b) $P = N + 3Q$
(c) $S = N - 2T$
(d) $H = \dfrac{N}{2} + 3P$
(e) $B = \dfrac{N + 2Y}{5}$
(f) $P = \dfrac{N - 5Q}{7}$

E5 Rearrange each formula so that the letter shown in red is the subject.

(a) $v = at + \mathbf{u}$
(b) $y = \dfrac{\mathbf{x} - 2c}{a}$
(c) $h = \mathbf{k} - lm$

What progress have you made?

Statement

I can form equations from a formula and solve them.

I can rearrange formulas.

Evidence

1 Pat's Pizzas use this rule to work out the cost of a pizza in pence (C)
$$C = 25n + 350$$
where n is the number of toppings.

(a) A pizza costs 475p. Form an equation and solve it to find the number of toppings on this pizza.

(b) Rearrange the formula to make n the subject.

(c) Use your formula to find n when $C = 425$.

2 Rearrange each formula so that the letter shown in red is the subject.

(a) $p = 2\mathbf{n} + 5$
(b) $h = 3(\mathbf{k} + 1)$
(c) $Y = \dfrac{\mathbf{X} - 3}{4}$
(d) $Y = \dfrac{\mathbf{X}}{4} - 3$
(e) $t = 5\mathbf{s} + p$
(f) $g = \mathbf{h} - 2k$

Review 1

1 This cube is hanging from a string fixed at the **middle** of an edge.
It is lowered into water.

Which of these are cross-sections for the cube at water level?

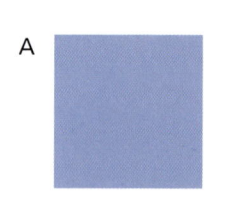

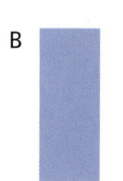

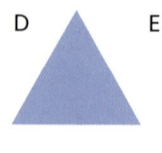

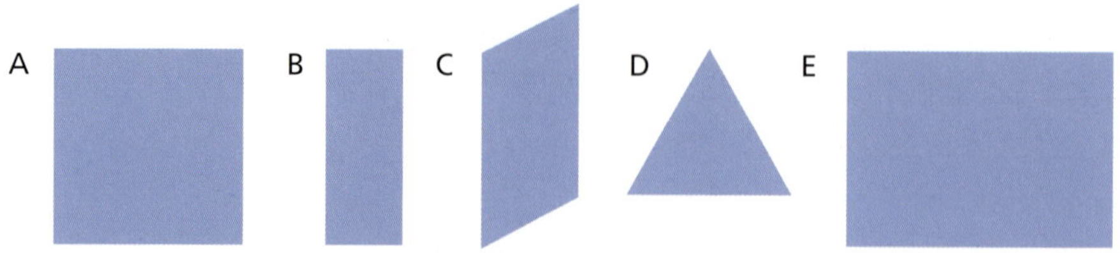

A   B   C   D   E

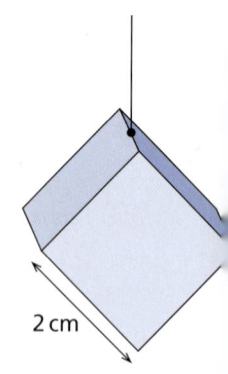

2 cm

2 Multiply out the brackets in each of these.

(a) $5(3n - 1)$ (b) $x(x + 5)$ (c) $y(1 - y)$

3 Copy and complete these addition walls.   (a)   (b)

The fraction on each brick is
found by adding the fractions
on the two bricks below.

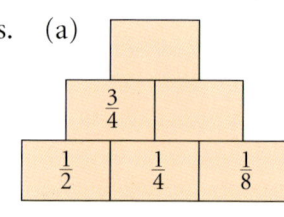

(a) bricks: $\frac{3}{4}$ on top of $\frac{1}{2}, \frac{1}{4}, \frac{1}{8}$

(b) bricks: $\frac{7}{10}$ on top of $\frac{1}{5}, \_, \frac{1}{3}$

4 Which of the stories matches the speed graph below?

A I cycled down a hill getting faster and faster. Then I freewheeled getting slower and slower until I stopped for a drink. Then I set off down the lane travelling at a steady speed.

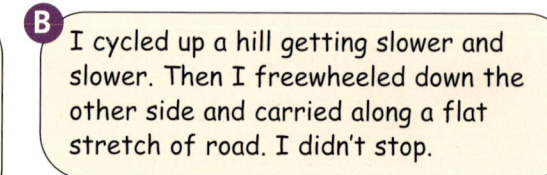

B I cycled up a hill getting slower and slower. Then I freewheeled down the other side and carried along a flat stretch of road. I didn't stop.

C I cycled up a hill getting slower and slower. Then I freewheeled down the other side and stopped at the bottom for a drink. I finished my journey along a flat stretch of road.

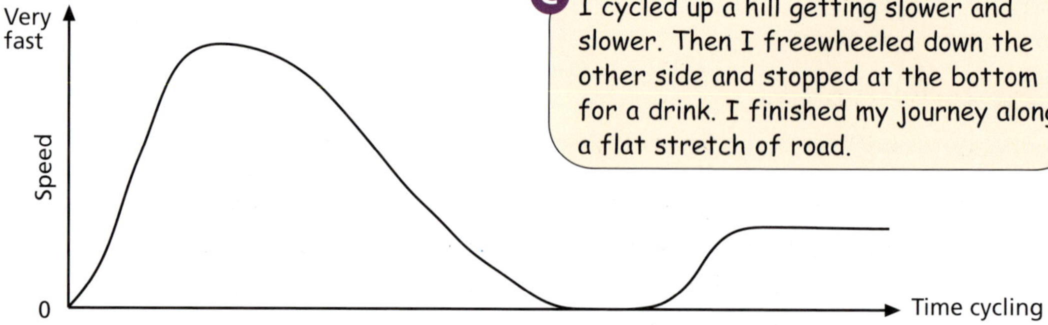

5 Work these out, simplifying where possible.

(a) $\frac{1}{2} \times \frac{1}{4}$ (b) $(\frac{1}{3})^2$ (c) $\frac{2}{5} \times \frac{1}{4}$ (d) $\frac{2}{3} \times \frac{3}{8}$

6 A regular pentagon has a perimeter of 10*n* units.
  Find an expression for the length of one edge.

7 Sometimes Equi's restaurant arrange their standard rectangular tables in a line like this.

  (a) If they use 5 tables in a line, how many chairs do they need?
  (b) A formula that tells you the number of chairs (*c*) when you know the number of tables (*t*) is
       $c = 2(3t + 1)$
       Which of the formulas below is equivalent to this?

       | $c = 5t + 1$ | $c = 6t + 1$ | $c = 3t + 2$ | $c = 6t + 2$ |

  (c) (i) Rearrange your formula to give a formula for *t* in terms of *c*.
      (ii) Use your formula to find the number of tables you would need to seat 56 people.

8 Jo ate $\frac{1}{4}$ of a bag of sweets.
  This is what she had left.
  How many sweets were in the whole bag?

9 Rearrange these to find a formula for *n* in terms of *p* each time.
  (a) $p = \frac{n-1}{6}$
  (b) $p = 2(n + 5)$
  (c) $p = \frac{n}{4} + 1$

10 The end of this prism is a right-angled triangle.
   (a) Work out the volume of this prism.
   (b) Calculate the surface area.
   (c) How many planes of symmetry does the prism have?

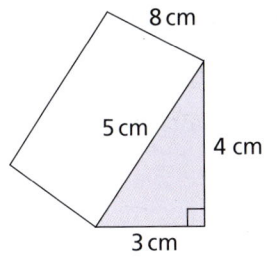

   8 cm
   5 cm
   4 cm
   3 cm

11 Factorise each of these.
   (a) $6n + 12$
   (b) $6n - 9$
   (c) $6n + n^2$

12 Pete's garden is mainly grass, but he also has a pond and a flower bed.
   Grass makes up $\frac{3}{4}$ of the garden and the rest is pond or flower bed.
   The pond takes up $\frac{1}{3}$ of the rest of the garden.
   The area of flower bed is 8 m².
   What is the area of grass?

6 Circumference of a circle

> This work will help you
> ◆ calculate a circle's circumference from its diameter or radius
> ◆ calculate its diameter or radius from its circumference
> ◆ solve problems that involve these measurements

A How many times?

The **circumference** of a circle is the distance all round it.

Make a 'circumference strip' for a cylindrical object.

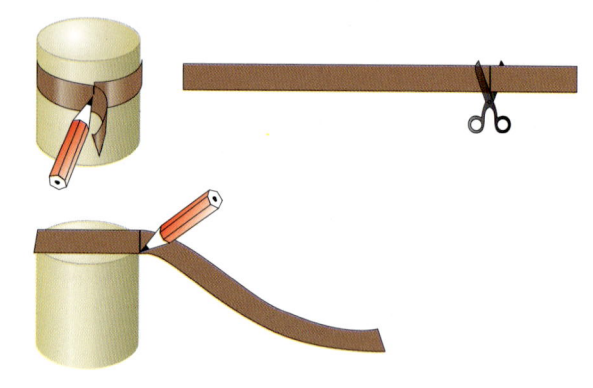

Mark the diameter on it.

Fold to see how many times the diameter goes into the circumference.

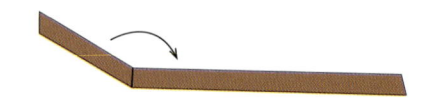

What do you find?

Is it true for any sized cylindrical object?

A1 What is the circumference of each of these, roughly?

(a) Peaches — Diameter = 7 cm

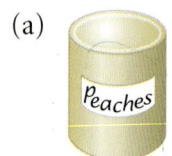

(b) Shoe Polish — Diameter = 10 cm

(c) Diameter = 6 cm

A2 Roughly how much sticky tape is needed to go once round the curved part of each of these parcels?

(a) 15 cm

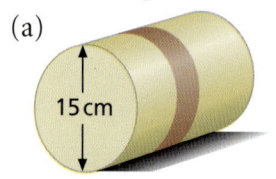

(b) 8 cm

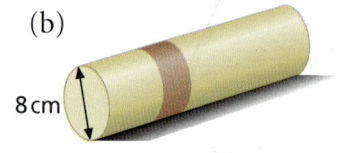

(c) 23 cm

A3 What is the circumference of each of these circles, roughly?

(a) 4 cm   (b) 5 cm   (c) 2.5 cm

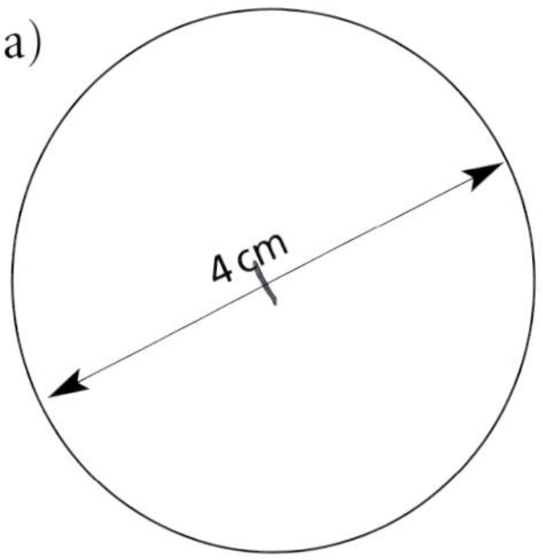

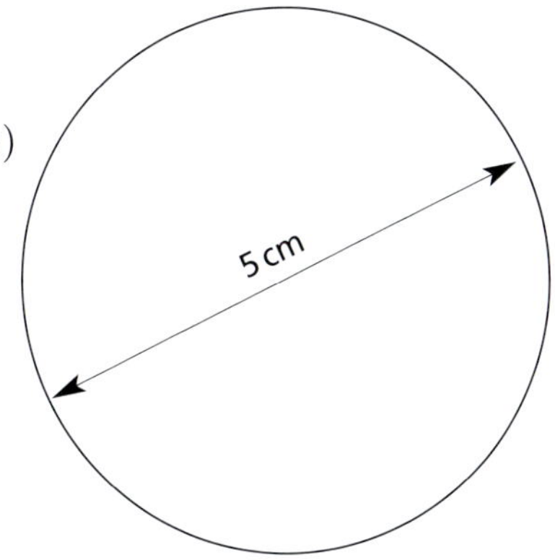

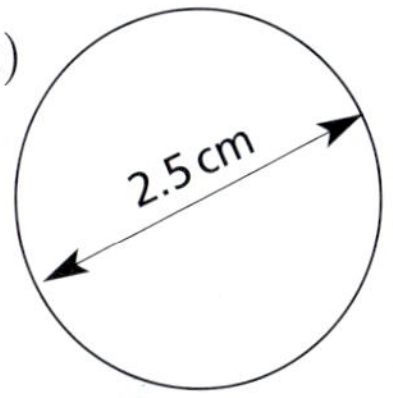

A4 A wedding ring has a diameter of 1.9 cm.
Roughly how long would the ring be if it was cut and straightened out?

A5 A cake tin has to be lined with greaseproof paper.
The diameter of the tin is 22 cm.
Roughly how long does the piece of paper have to be?

A6 This is a design for an earring made from silver wire.
About how much wire will be needed?

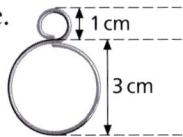

1 cm
3 cm

A7 A reel of cotton has 1000 turns on it.
The reel has a diameter of 3 cm.
Roughly how much cotton is on the reel?

B Finding the diameter and radius

I could find the tree's diameter by chopping it down, or I could get a rough answer working backwards.

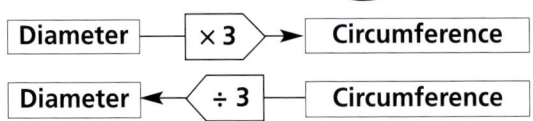

Diameter — × 3 → Circumference

Diameter ← ÷ 3 — Circumference

B1 The circumference of the trunk of a small tree is 60 cm.
What is its diameter, roughly?

B2 The circumference of a mug is 21 cm.
What is its diameter, roughly?

B3 The distance round a large pipe is 48 cm.
What is its diameter, roughly?

B4 The curved surface of a can of beans is made from this sheet of metal.

What is the diameter of the can, roughly?

B5 A farmer buys 35 m of wire fencing.
He wants to make a circular sheep pen with it.
What will the diameter of the pen be, roughly?

B6 The distance round a hose-pipe is 4.5 cm. What is its diameter roughly?

B7 At Avebury, in Wiltshire, there is a prehistoric circle of stones.
The distance all round the circle is 1150 m.
What is the diameter of the circle, roughly?

B8 The distance round the equator is about 25 000 miles.
What is the diameter of the Earth, roughly?

Remember, the **radius** of a circle is half its diameter.

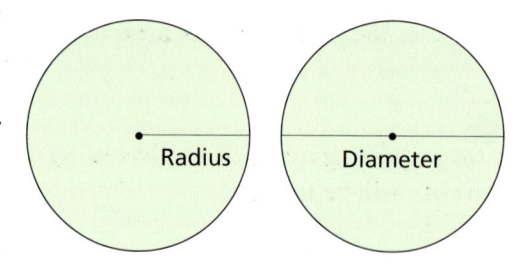

B9 A circle has a circumference of 42 cm. What is its **radius**, roughly?

B10 A piece of wire 6 metres long is bent into the shape of a circle.
What is the radius of the circle, roughly?

B11 The 'Inner Circle' is a circular road in Regent's Park, London.
A man paces the distance round it and finds it is 1140 paces.
How many paces would he have to go to reach the centre of the gardens inside the Inner Circle?

B12 10 people stand in a circle holding hands like this.

(a) Roughly how far apart do you think the people will be?

(b) What will the circumference of the circle be?

(c) What will its diameter be?

B13 Silbury Hill in Wiltshire is a prehistoric mound with a circular base.
The distance round the bottom of the hill is 530 m.

In 1968, archaeologists dug a tunnel at the base, to the centre of the hill.
Roughly how long was their tunnel?

C Becoming more accurate

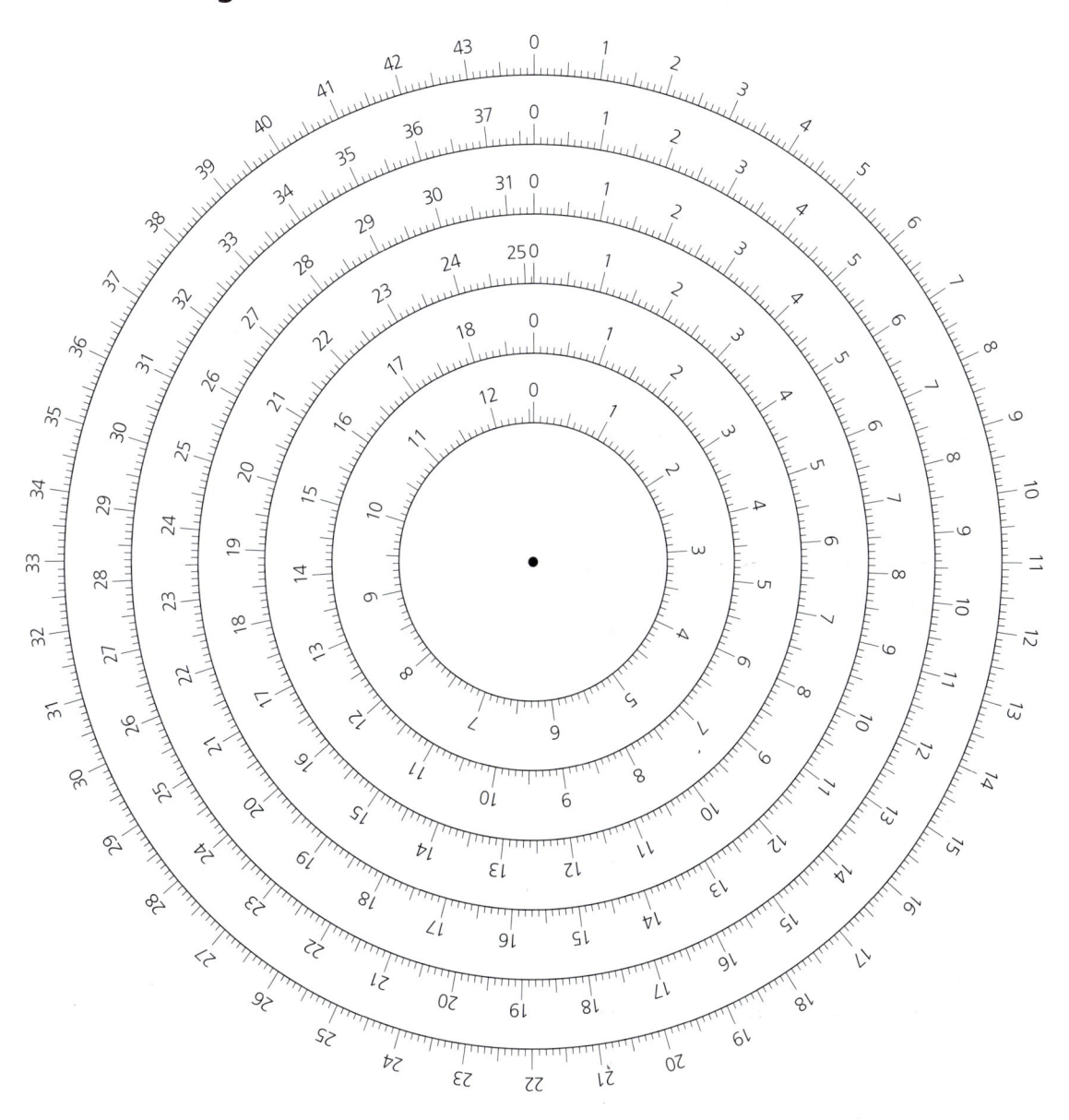

C1 The circumferences of these circles are marked off in centimetres and tenths of a centimetre.

Copy this table and fill in the circumference and diameter for all six circles.

For each circle, find a number that you multiply the diameter by to get the circumference.

Write each 'multiplier' in the table.

Diameter	×?	Circumference

Your 'multipliers' in C1 should be between 3.13 and 3.15.

The exact number to multiply by is called π (pronounced 'pie').

If you press the π key on your calculator, you get a value with many decimal places, far more accurate than you need for everyday purposes.

3.141592654

This stands for 4 billionths of a unit – a very small amount!

Formulas for circumference

Let r be the radius of a circle,
 d the diameter,
 C the circumference.

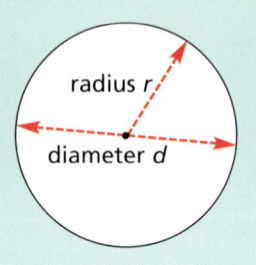

Then $C = \pi d$

Also, because $d = 2r$, it follows that

$$C = \pi \times 2r$$

which is written

$$C = 2\pi r$$

Use the π key on your calculator for the following questions.

C2 Calculate the circumference of the each of these circles. Round each answer to the nearest 0.1 cm.

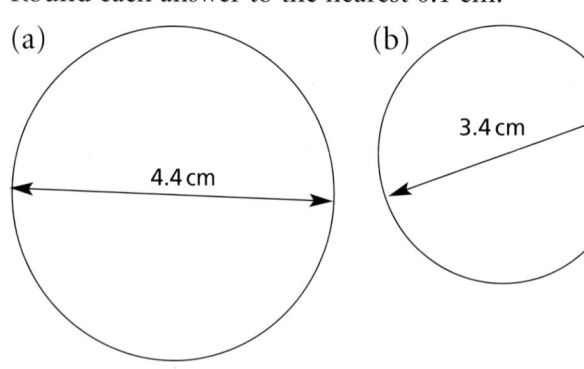

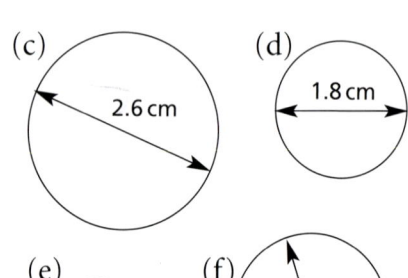

(a) 4.4 cm

(b) 3.4 cm

(c) 2.6 cm

(d) 1.8 cm

(e) 0.9 cm

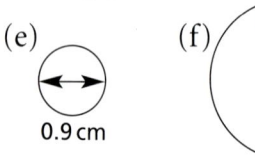

(f) 2.1 cm

C3 The radius of this circle is 1.3 cm.

(a) Write down the diameter.

(b) Calculate the circumference, to the nearest 0.1 cm.

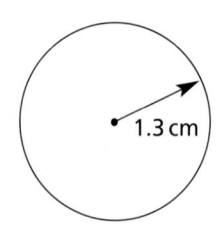

1.3 cm

C4 Calculate, to the nearest 0.1 cm, the circumference of a circle whose radius is

(a) 1.9 cm   (b) 1.4 cm   (c) 3.6 cm   (d) 0.8 cm   (e) 5.9 cm

C5 Calculate, to the nearest 0.1 cm, the circumference of a circle with

(a) diameter 6.4 cm (b) radius 12.8 cm (c) diameter 10.5 cm
(d) radius 23.2 cm (e) diameter 15.2 cm (f) radius 15.2 cm

C6 For each of these circles, measure and record the diameter and then work out the circumference to the nearest 0.1 cm.

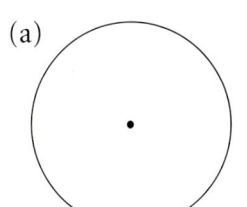

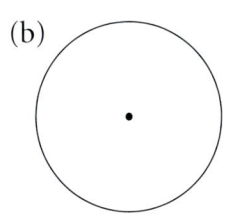

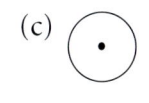

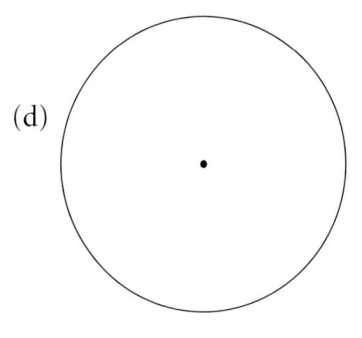

(a)   (b)   (c)   (d)

Towards greater accuracy

A round number

The Bible describes a 'sea of cast metal' in Solomon's temple, with these dimensions.

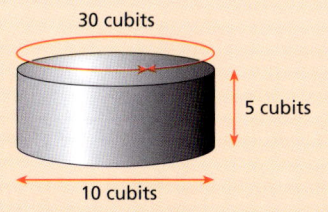

30 cubits
5 cubits
10 cubits

(A cubit is about half a metre.)

This suggests π is 3, the rough value you used earlier.

Egypt and Babylon

Nearly 4000 years ago the Egyptians used a value of 3.16.
The ancient Babylonians used 3.125.
You can get this degree of accuracy from measuring, as in section C.

π from polygons

For hundreds of years people found more accurate values of π by calculating the perimeter of polygons inside and outside a circle, doubling the number of sides each time.

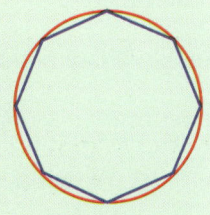

In 1593 a Dutchman, Adriaen Romanus, found π to 15 decimal places, by considering a polygon with over 100 million sides!

Billions of decimal places

With the help of computers, π has been calculated to many billions of decimal places.

Mathematicians have proved that, however many decimal places you calculate, you will never get the exact value of π.

3.141592653589793238462643383
2795028841971693993751058209
74944592307816406286208998628
03482534211706798214808651328
23066470938446095505822317253
59408128481117450284102701938
52110555964462294895493038196
442881097566593344612847564823

D Calculating diameter and radius

The formula $C = \pi d$ can be shown as a flow diagram.

Reversing the diagram leads to a formula for finding d.

$$d = \frac{C}{\pi}$$

The formula $C = 2\pi r$ can be shown as a flow diagram.

Reversing the diagram leads to a formula for finding r.

$$r = \frac{C}{2\pi}$$

D1 Calculate, to the nearest 0.1 cm, the diameters of circles that have these circumferences.
(a) 4.9 cm (b) 7.2 cm (c) 8.5 cm (d) 9.8 cm

D2 Sean wanted to calculate the radius of a circle of circumference 58 cm.
Here is the key sequence he used on his calculator.   5 8 ÷ 2 × π =
Is this sequence correct? If not, what should it be?

D3 Calculate, to the nearest 0.1 cm, the radius of a circle whose circumference is
(a) 67 cm (b) 146 cm (c) 44.8 cm (d) 106.2 cm (e) 9.4 cm

D4 Petra has a piece of silver wire 20.0 cm long.
She wants to make it into a circular bracelet. What diameter will it have?

D5 These are descriptions of three circular plates.

Plate P has radius 12.4 cm.   Plate Q has diameter 20.4 cm.
Plate R has circumference 66.6 cm.

(a) Which plate is biggest? (b) Which is smallest?

D6 An iron strip 5.30 metres long is made into a circular band to go round a wagon wheel.
What is the diameter of the wheel?

What progress have you made?

Statement

I know what the words 'diameter' and 'radius' mean.

I can calculate a circumference from a diameter or a radius.

I can calculate a radius or a diameter from a circumference.

Evidence

1 Measure the diameter and radius of this circle.

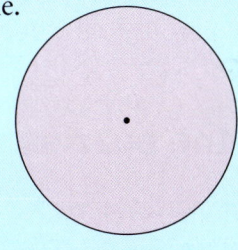

2 What is the circumference of a circle with diameter 2000 metres?

3 What is the circumference of a circle with radius 2.8 cm?

4 A circular cycle track has an inside radius of 100 m. The track is 8 m wide.

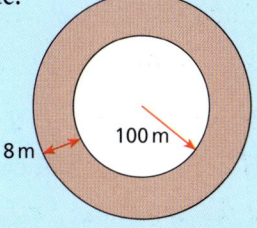

How much further does a cyclist go in one lap if she keeps to the outside of the track rather than the inside?
(Give your answer to the nearest metre.)

5 What is the diameter of a pole with circumference 37 cm?

6 What is the radius of a circular pond with a path round it 100 metres long?

7 Clue-sharing

> This work will help you
> - revise factors, multiples, prime numbers, square numbers and powers
> - solve problems by thinking logically

A Solving the puzzles

For groups, pairs or individual work

Each puzzle uses one of these boards.

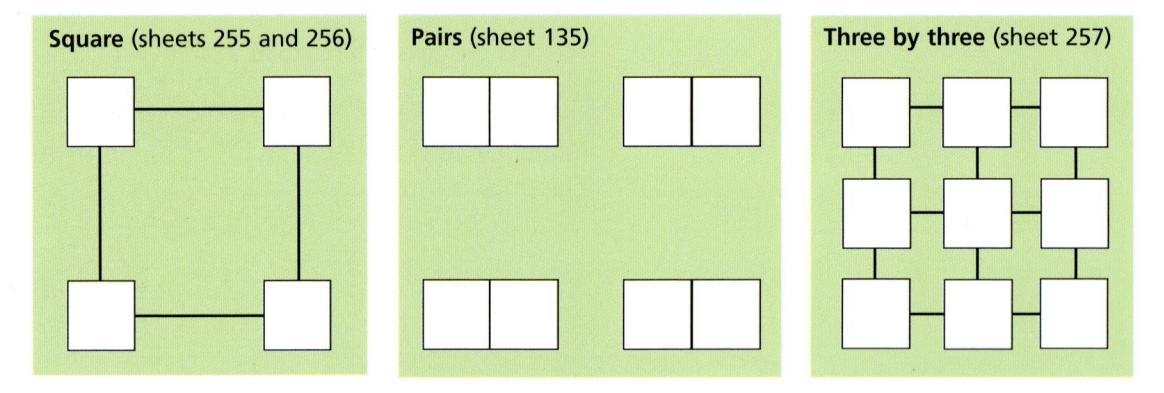

Square (sheets 255 and 256)

Pairs (sheet 135)

Three by three (sheet 257)

- Cut out the clue cards and number cards for the puzzle you are doing.
- Use the clues to decide which numbers go on the board and where they go.

You will need to know the **mathematical** meaning of these words. (Product) (Sum)

B Inventing and trying out your own puzzles

- Choose one of the boards above or design one of your own.
- Put numbers on the board.
- Make up clues to fit the numbers on the board.
- Give your board, your clues and the numbers for your puzzle to another pupil or pair of pupils and challenge them to solve it. You must not give them hints. If the puzzle 'works' they should be able to solve it using your written clues alone.

8 Enlargement

This work will help you
- enlarge a shape from a centre
- solve enlargement problems

A Enlargement from a centre

This way of enlarging a shape uses a **centre of enlargement**.
It is sometimes called the 'ray method'.

Draw lines from O, the centre of enlargement, through the vertices.

Measure the distance from O to each vertex.

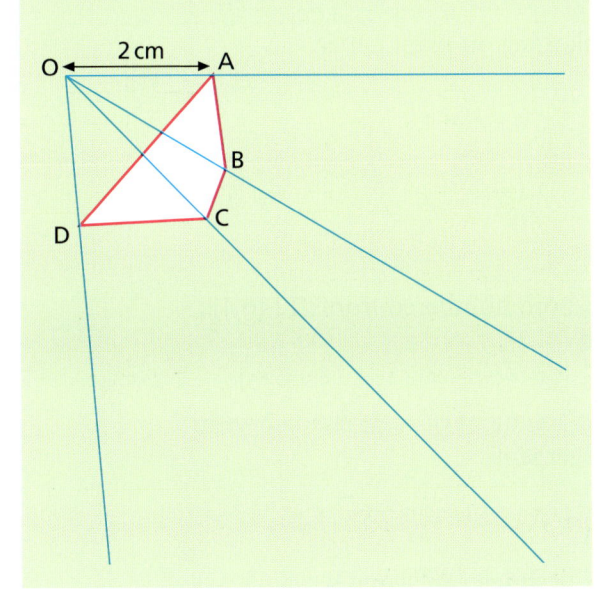

If the scale factor is 3, multiply the distances from the centre by 3.

Measure the new distances along the lines starting from O, and mark the vertices of the image.

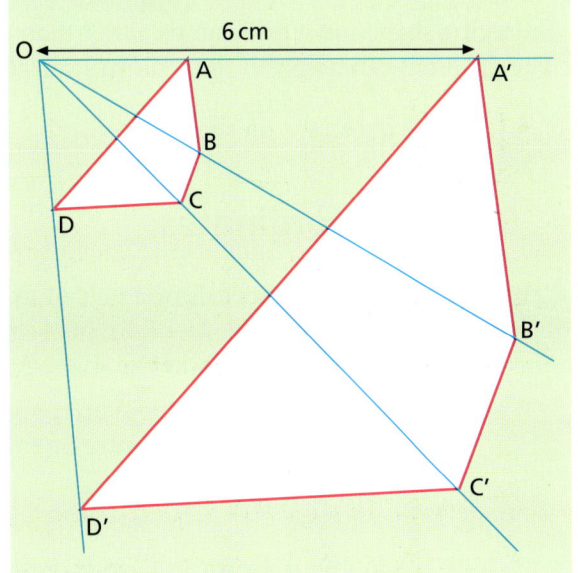

- Check that the lengths OB, OC and OD have been multiplied by 3.
- What has happened to
  - the length of the sides?
  - the angles?
  - the perimeter?

A1 Enlarge the shapes on sheet 258.
Use the centres of enlargement and scale factors given on the sheet.

A2 The shape PQRS has been enlarged by scale factor 2 to give P′Q′R′S′.

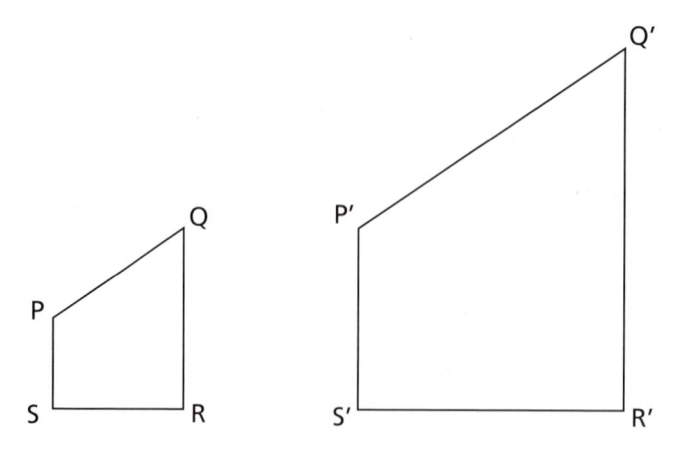

(a) Angle QPS = 124°.
   Write down the size of angle Q′P′S′.

(b) The perimeter of PQRS is 7.55 cm.
   Calculate the perimeter of P′Q′R′S′.

Enlarging shapes on a grid can be easier than enlarging on plain paper.
The gridlines can help you to be more accurate.

A3 Do the questions on sheet 259.

B Using coordinates

B1 (a) On squared paper draw a pair of axes, both numbered from ⁻12 to 12.
   Plot the points (2, 2), (2, 7), (4, 4) and (4, 2).
   Join them up and label the shape A.

(b) Draw the image of shape A after an enlargement of scale factor 2 with
   the point (4, 11) as the centre of enlargement.
   Label the image B.

(c) (i) Measure the perimeter of shape A.

   (ii) Without measuring, work out the perimeter of shape B.

(d) Draw the image of shape A after an enlargement of scale factor 4 with
   the point (6, 6) as the centre of enlargement.
   Label the image C.

(e) Shape C is the image of shape B after an enlargement.

   (i) What is the scale factor of this enlargement?

   (ii) Find the coordinates of the centre of enlargement.

B2 Find the centres of enlargement and scale factors for the enlargements on sheet 260.

B3 Draw a pair of axes, both numbered from ⁻8 to 8.

(a) (i) Plot and label the points A (0, 2), B (3, 2) and C (3, 4). Join them up.

(ii) Enlarge ABC by a scale factor 2 with centre of enlargement (0, 0). Label the image points A′, B′ and C′.

(iii) Copy and complete this table.

Coordinates of original shape	Coordinates of image
A (0, 2)	
B (3, 2)	
C (3, 4)	

(b) Repeat (a) but use the points P (⁻1, 3), Q (⁻1, ⁻3), R (⁻4, ⁻3) and S (⁻4, 1).

(c) Answer this question without drawing.

The points X (3, 1), Y (4, ⁻1) and Z (0, ⁻4) are the vertices of shape XYZ. The shape is enlarged by a scale factor of 2 with centre of enlargement (0, 0). Work out the coordinates of the vertices of the image.

B4 Draw a pair of axes, both numbered from ⁻9 to 9.

(a) (i) Plot and label the points A (3, 3), B (⁻2, 2), C (⁻1, ⁻1) and D (1, ⁻2). Join them up.

(ii) Enlarge ABCD by a scale factor 3 with centre of enlargement (0, 0). Label the image points A′, B′, C′ and D′.

(iii) Copy and complete this table.

Coordinates of original shape	Coordinates of image
A (3, 3)	
B (⁻2, 2)	
C (⁻1, ⁻1)	
D (1, ⁻2)	

(b) The points F (3, 1), G (4, 5) and H (2, ⁻5) are the vertices of shape FGH. The shape is enlarged by a scale factor of 3 with centre of enlargement (0, 0). Work out the coordinates of the vertices of the image.

B5 A shape has vertices P (2, 1), Q (3, 7) and R (⁻5, 2). It is enlarged by a scale factor of 4 with centre of enlargement (0, 0). Can you work out the coordinates of the vertices of the image?

B6 A shape is enlarged by a scale factor of 5 with centre of enlargement (0, 0).
The **image** shape has vertices K′ (10, 15), L′ (20, 5) and M′ (−5, −30).
Can you work out the coordinates of the original vertices K, L and M?

***B7** Draw a pair of axes, both numbered from 0 to 8.
  (a) Plot and label the points A (2, 1), B (2, 4), C (4, 3) and D (4, 2).
      Join them up.
  (b) Enlarge ABCD by a scale factor 2 with centre of enlargement (0, 2).
      Label the image points A′, B′, C′ and D′.
  (c) Make a table to show the coordinates of A, B, C and D with their images.
  (d) Can you find a rule that links the coordinates of a point with its image after an enlargement of scale factor 2 with (0, 2) as the centre of enlargement?

C Scale factors

- How can you find the scale factor of this enlargement?

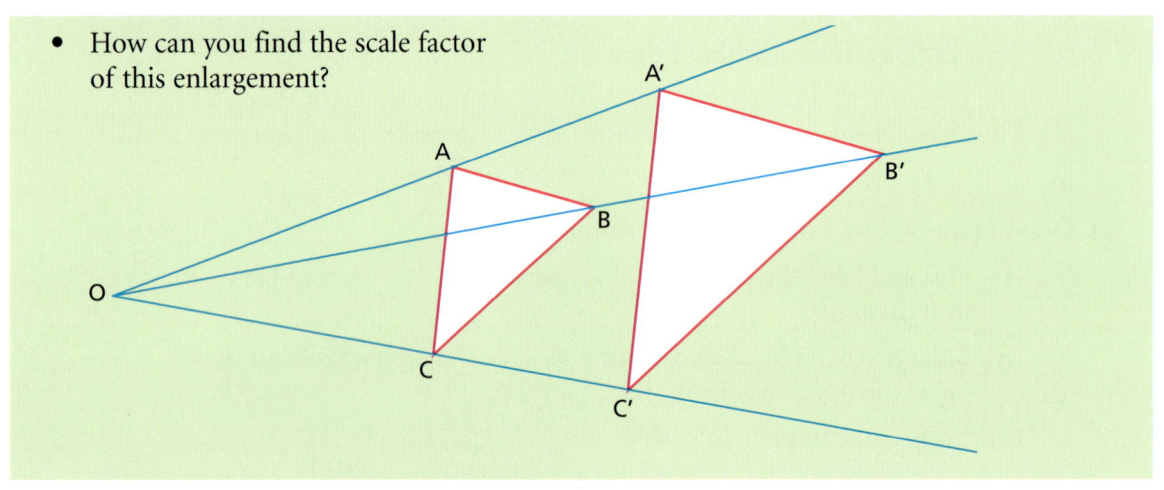

C1 Shape P′Q′R′S′ is an enlargement of PQRS.

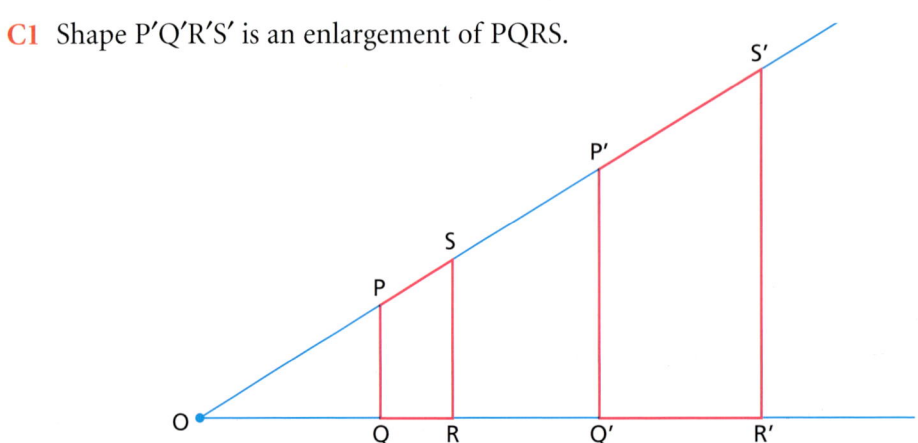

  (a) Measure lines PQ and P′Q′.
  (b) Use these lengths to work out the scale factor of this enlargement.

C2 In each diagram below, a triangle has been enlarged using the 'ray method'.
Only part of each enlargement is shown.
Make suitable measurements and work out each scale factor.

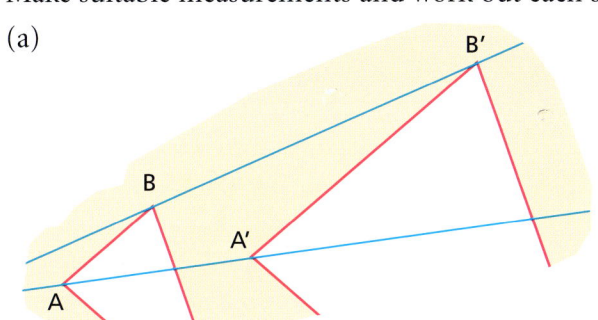

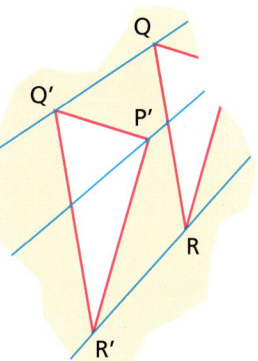

(a)

(b)

(c)

(d)

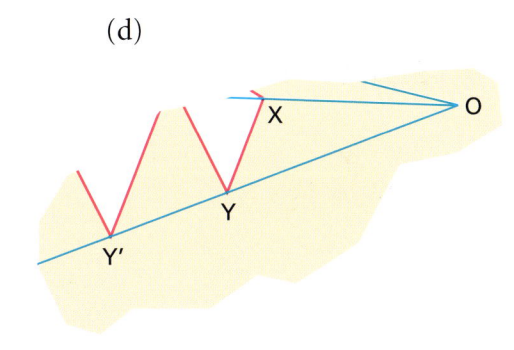

C3 A triangle PQR has been enlarged using the 'ray method'.
Part of the enlargement with some lengths is shown below.
Work out the scale factor of the enlargement.

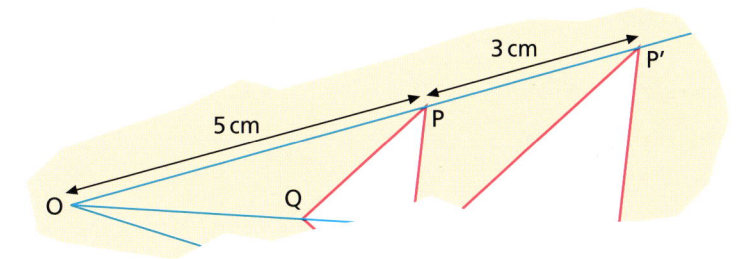

C4 Triangle A'B'C' is the image of ABC after an enlargement with centre O.
Without measuring, work out the value of

(a) $\dfrac{B'C'}{BC}$ (b) $\dfrac{OA'}{OA}$

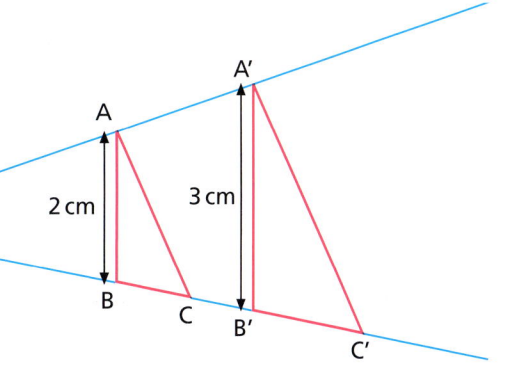

What progress have you made?

Statement

I can enlarge a shape given the scale factor and centre of enlargement.

Evidence

1 Draw a pair of axes, both numbered from −12 to 12.

  (a) Plot the points (4, 2), (4, 5), (6, 7) and (8, 5). Join them up and label the shape P.

  (b) Draw the image of shape P after an enlargement of scale factor 3 with the point (6, 9) as the centre of enlargement. Label the image Q.

  (c) Draw the image of shape P after an enlargement of scale factor 2 with the point (12, 8) as the centre of enlargement. Label the image R.

  (d) Shape Q is the image of shape R after an enlargement.

   (i) Find the scale factor of this enlargement.

   (ii) What are the coordinates of the centre of enlargement?

I can work out the coordinates of a point after an enlargement with centre (0, 0).

2 A shape has vertices A (6, 1), B (−1, 4) and C (0, 7). It is enlarged by scale factor 2 with centre of enlargement (0, 0).

Without drawing, work out the coordinates of the vertices of the image.

I can work with scale factors.

3 Triangle X′Y′Z′ is the image of XYZ after an enlargement with centre O.

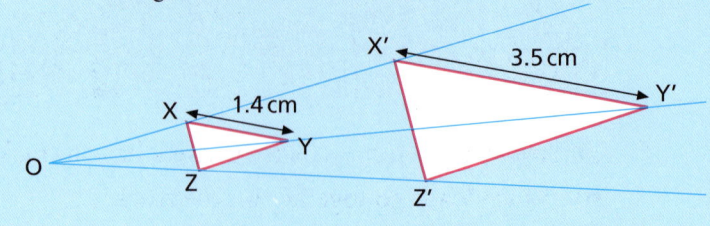

  (a) Find the value of $\dfrac{OX'}{OX}$.

  (b) The perimeter of triangle XYZ is 3.3 cm. Work out the perimeter of X′Y′Z′.

  (c) Angle ZXY = 65°. What is the size of angle Z′X′Y′?

9 Over to you

This is a collection of problems.
Working on them will help you
- think about how you might try to solve a problem
- try out ideas and change your approach if necessary
- explain your reasoning

1   The duchess dies.
    She leaves her collection of vases to her three children.
    She wants each of them to have an equal share of her collection.

    The value of each vase is shown.

    £46   £23   £42   £27   £19   £38

    (a) How should the children split the collection of vases?
    (b) Jane inherits the vase worth £46.
        Which other vase will she inherit?

2   These loads have to be carried.
    The truck will carry up to 1000 kg at a time.

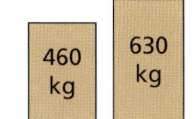

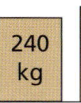

    460 kg   630 kg   270 kg   150 kg   240 kg   380 kg   290 kg   140 kg   100 kg   310 kg

    (a) How many journeys will the truck have to make?
    (b) Which loads go together on the truck?

3   How much do you pour from each jug A, B, C, D, into E to make the amounts in every jug the same?

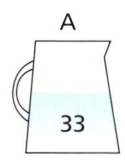

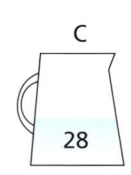

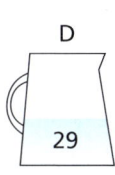

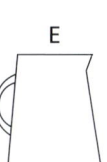

    A 33   B 45   C 28   D 29   E

4 Which pair of loads have to be swapped to make the scales balance?

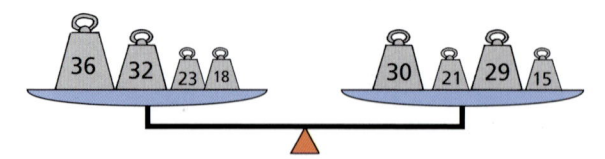

Left: 36, 32, 23, 18   Right: 30, 21, 29, 15

5 Split £47 between three people A, B and C so that A gets £10 more than B and B gets £5 more than C.

6 Which weight do you move from left to right to right to make the scales balance?

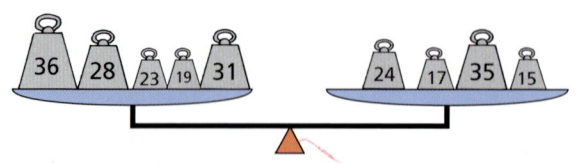

Left: 36, 28, 23, 19, 31   Right: 24, 17, 35, 15

7 These sheds are equally spaced.
What is the size of the gap between each shed and the next?

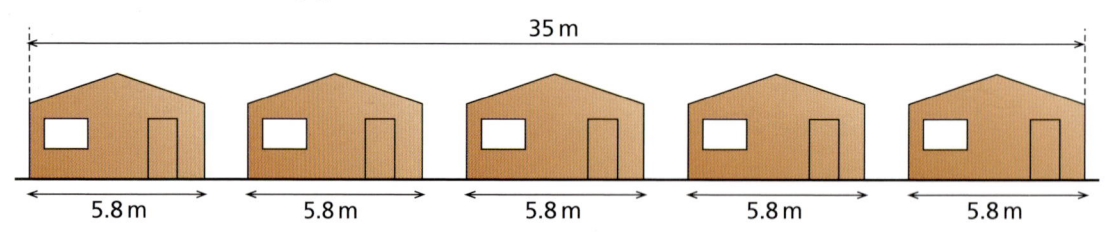

Total: 35 m. Spacings shown: 5.8 m, 5.8 m, 5.8 m, 5.8 m, 5.8 m

8 How can you share these jewels between three people so that each gets the same weight?

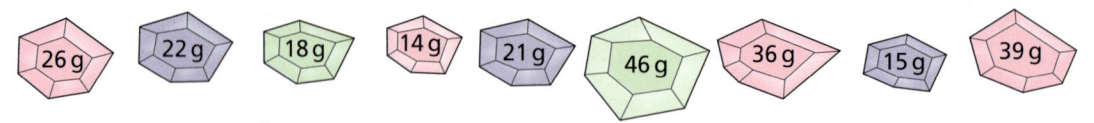

26 g, 22 g, 18 g, 14 g, 21 g, 46 g, 36 g, 15 g, 39 g

9 Work out the missing lengths.
(All the angles are right angles.)

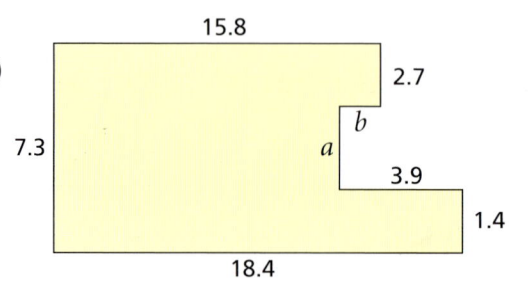

Given: 15.8, 2.7, b, a, 3.9, 1.4, 18.4, 7.3

10 Alan is 15 years older than Carol.
Bharat is 10 years older than Sadia.
Carol is 14 years older than Pat.
Bharat is 16 years older than Kevin.
Bharat is 7 years older than Carol.

Put the people in older of age oldest to youngest.

10 Straight-line graphs

This work will help you
- draw straight-line graphs from algebraic equations
- work out the gradient and *y*-intercept of a straight line
- use the gradient and *y*-intercept of a straight line to work out its equation

A Connections

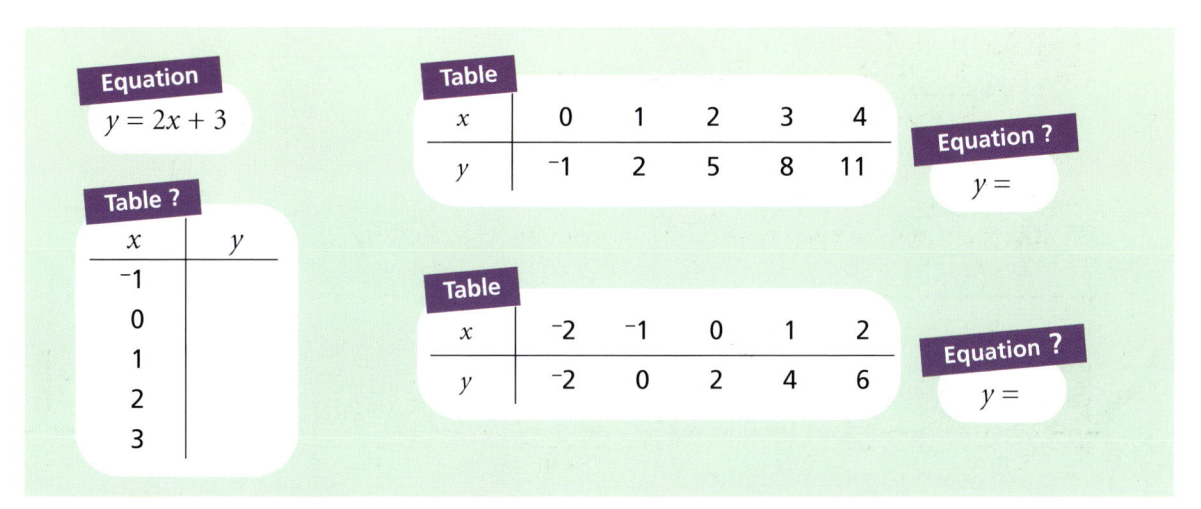

Equation
$y = 2x + 3$

Table ?

x	y
-1	
0	
1	
2	
3	

Table

x	0	1	2	3	4
y	-1	2	5	8	11

Equation ?
$y =$

Table

x	-2	-1	0	1	2
y	-2	0	2	4	6

Equation ?
$y =$

A1 Look at this table of values of *x* and *y*. Which of the equations below fits all the values in the table?

x	0	1	2	3	4	5
y	1	4	7	10	13	16

A $y = x + 1$ **B** $y = 2x + 1$ **C** $y = 4x$ **D** $y = 3x + 1$

A2 (a) Match each equation below to a table of values.

1 $y = x + 4$
2 $y = 3x$
3 $y = 2x + 1$
4 $y = 5x - 2$

P

x	0	1	2	3
y	4	6	8	10

R

x	0	2	4	6
y	-2	8	18	28

Q

x	y
1	3
2	5
3	7
4	9

S

x	y
-1	3
0	4
1	5
2	6

T

x	-2	-1	0	1
y	-6	-3	0	3

(b) One of the tables has no matching equation. Write down an equation that fits this table.

B From equation to graph

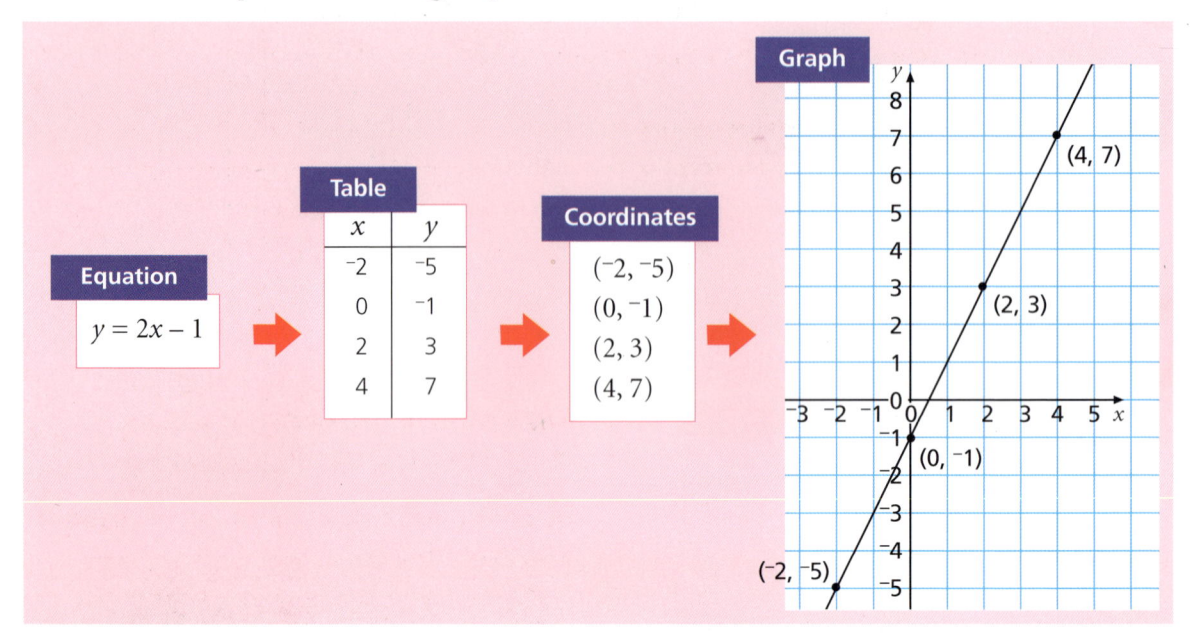

Equation: $y = 2x - 1$

Table:

x	y
-2	-5
0	-1
2	3
4	7

Coordinates: (-2, -5), (0, -1), (2, 3), (4, 7)

Graph: points (-2, -5), (0, -1), (2, 3), (4, 7) plotted on axes.

Plotting straight-line graphs

1. Find the coordinates of three or more points that fit the equation.
2. Think carefully about how long to make your axes.
3. Draw your axes and plot the points.
4. Check they lie in a straight line.
5. Draw the line and label it with its equation.

B1 Draw the graph of $y = 2x + 3$ like this.

(a) Copy and complete this table for $y = 2x + 3$.

x	y
-2	
0	
2	
4	

(b) Draw axes (x going from -2 to 4, and y going from -1 to 11 for this graph).

(c) Plot the points you found in (a). Check that they lie in a straight line.

(d) Draw and label the line.

B2 For each of these equations,
- find some points that fit the equation
- draw suitable axes on squared paper
- plot the points; draw and label the graph

(a) $y = x + 5$ (b) $y = x$ (c) $y = 4x + 3$ (d) $y = 3x - 4$

B3 (a) Write down the coordinates of five points that lie on the red line.

(b) The coordinates all fit one of these equations. Which one is it?

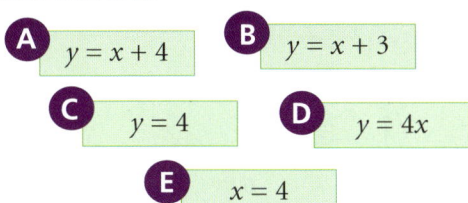

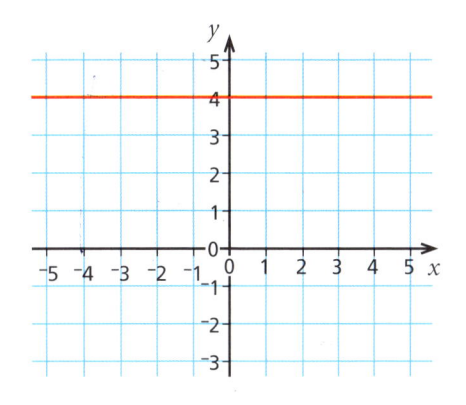

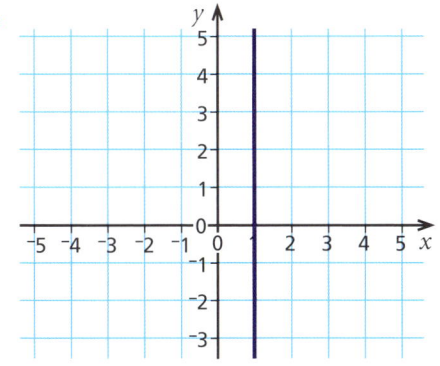

- **A** $y = x + 4$
- **B** $y = x + 3$
- **C** $y = 4$
- **D** $y = 4x$
- **E** $x = 4$

B4 (a) Write down the coordinates of five points that lie on the blue line.

(b) The coordinates all fit one of these equations. Which one is it?

- **A** $y = x$
- **B** $y = 1$
- **C** $y = 2x$
- **D** $y = x + 1$
- **E** $x = 1$

B5 Work out the equations of each of these four lines.

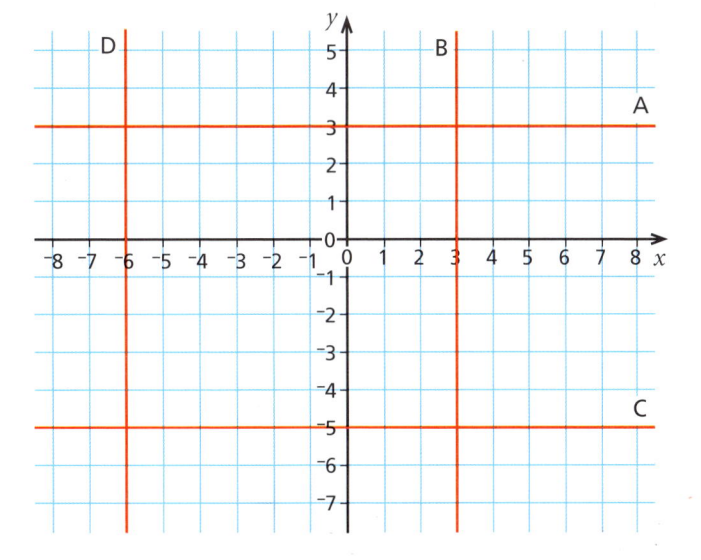

B6 On squared paper, draw a pair of axes numbered from ⁻5 to 5 along each axis.
Draw each of these lines on your axes.

(a) $y = x + 2$ (b) $y = 2$ (c) $x = ^-2$ (d) $y = x - 2$

65

C Gradient and intercept

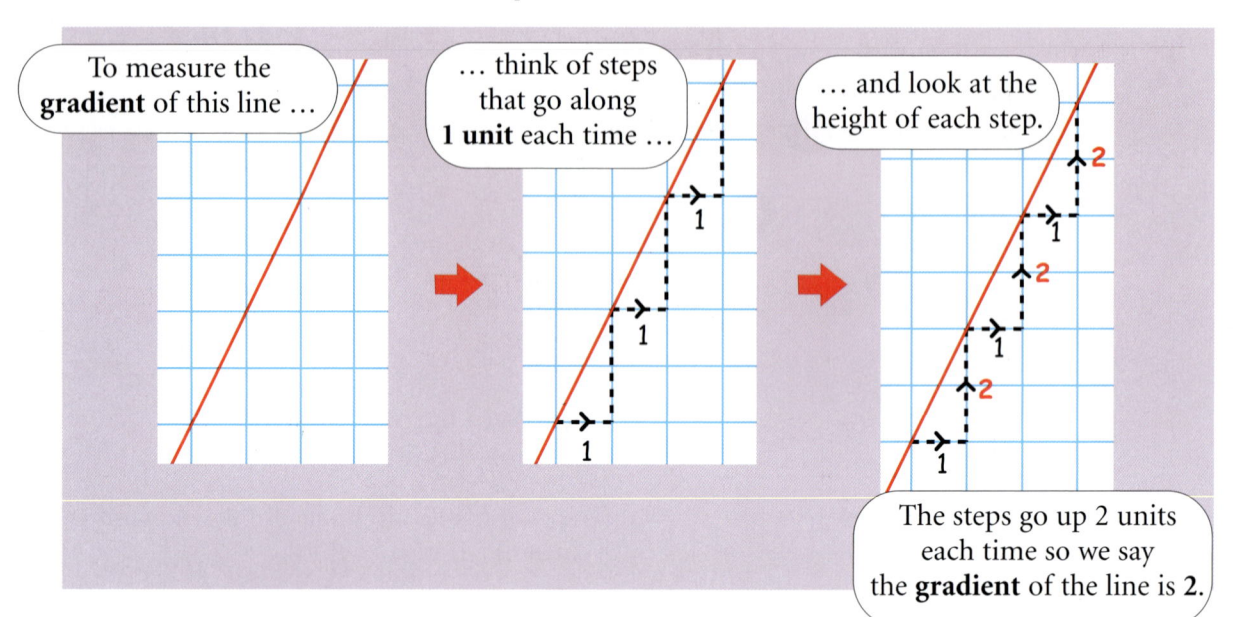

To measure the **gradient** of this line …

… think of steps that go along **1 unit** each time …

… and look at the height of each step.

The steps go up 2 units each time so we say the **gradient** of the line is **2**.

C1 Write down the gradient of each of these lines.

(a) (b) 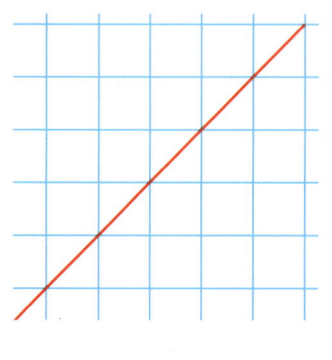 (c)

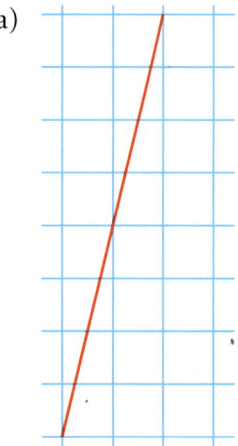

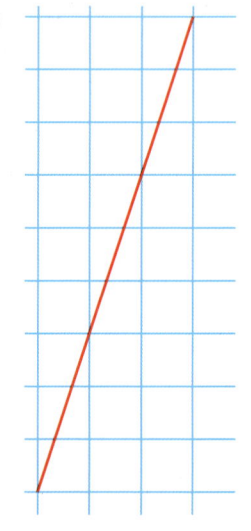

C2 Complete sheet 261.

C3 Complete sheet 262.

C4 Without drawing, give the gradient of the following lines.

(a) $y = 2x + 1$ (b) $y = 6x + 10$ (c) $y = 3x - 3$ (d) $y = x - 15$

C5 (a) Which equations below give lines with the same gradient?

A $y = 3x + 4$ B $y = 2x + 3$ C $y = 5x + 4$ D $y = 3x - 1$

(b) Which equations above give lines with the same y-intercept?

D From graph to equation

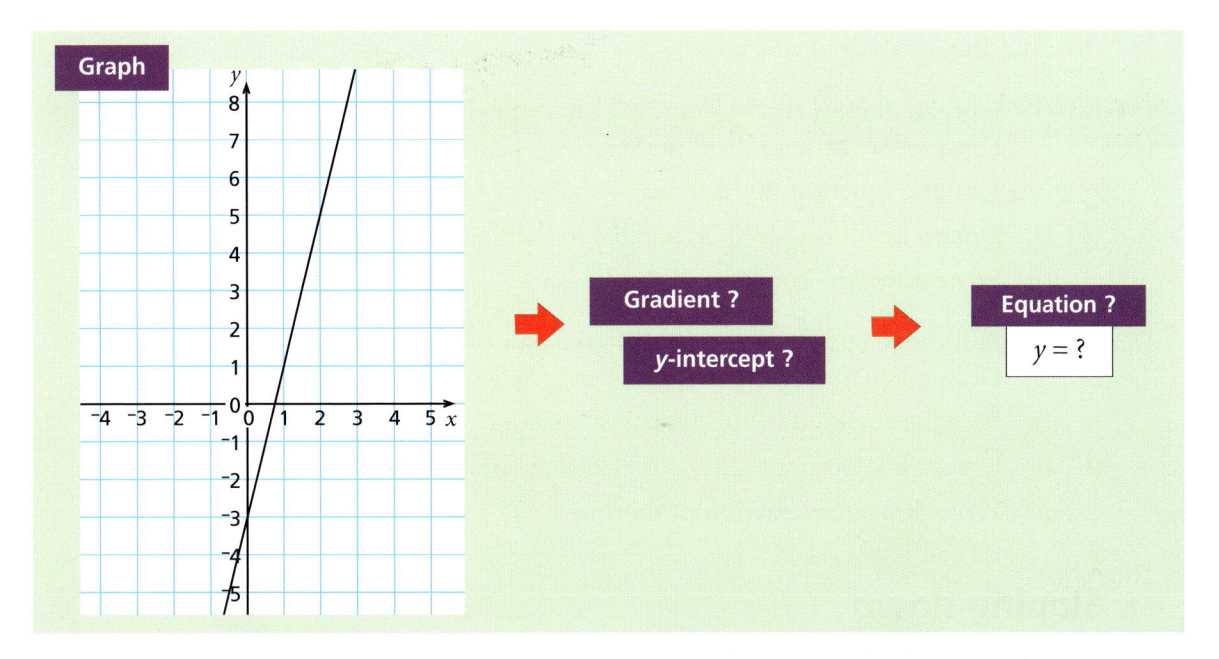

Graph → **Gradient ?** / **y-intercept ?** → **Equation ?** $y = ?$

D1 Look at this graph.
(a) What is the gradient of the straight line?
(b) What is its y-intercept?
(c) Write down its equation.

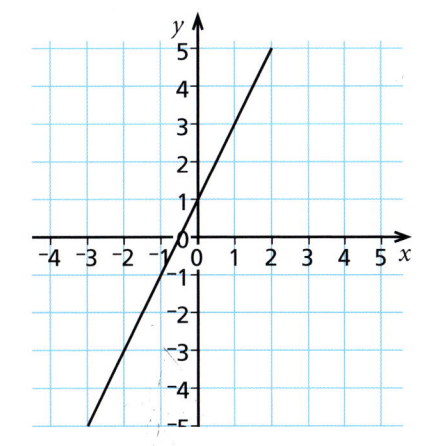

D2 Look at this graph.
(a) What is the gradient of the straight line?
(b) What is its y-intercept?
(c) Write down its equation.

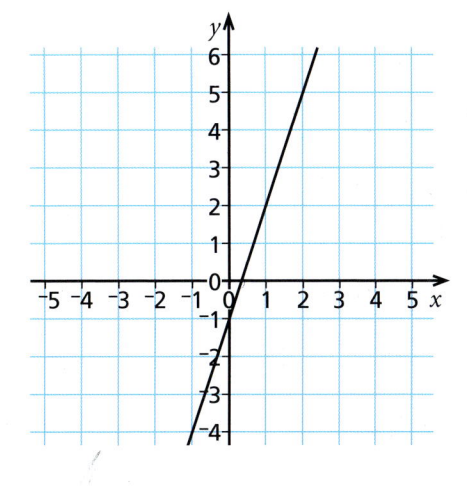

D3 (a) Plot the points (1, 4) and (3, 8) on a pair of axes.
Draw a line through these points.

(b) Find the equation of this line.

D4 (a) (i) Plot the points (⁻1, ⁻5) and (2, 7) on a pair of axes.
Draw a line through these points.

(ii) Find the equation of this line.

(b) (i) Draw a line through (0, 3) parallel to the first line.

(ii) Write down the equation of this line.

D5 (a) (i) Plot the points (1, 4) and (3, 10) on a pair of axes.
Draw a line through these points.

(ii) Find the equation of this line.

(b) (i) Draw a line through (2, 1) parallel to the first line.

(ii) Write down the equation of this line.

E Sloping down

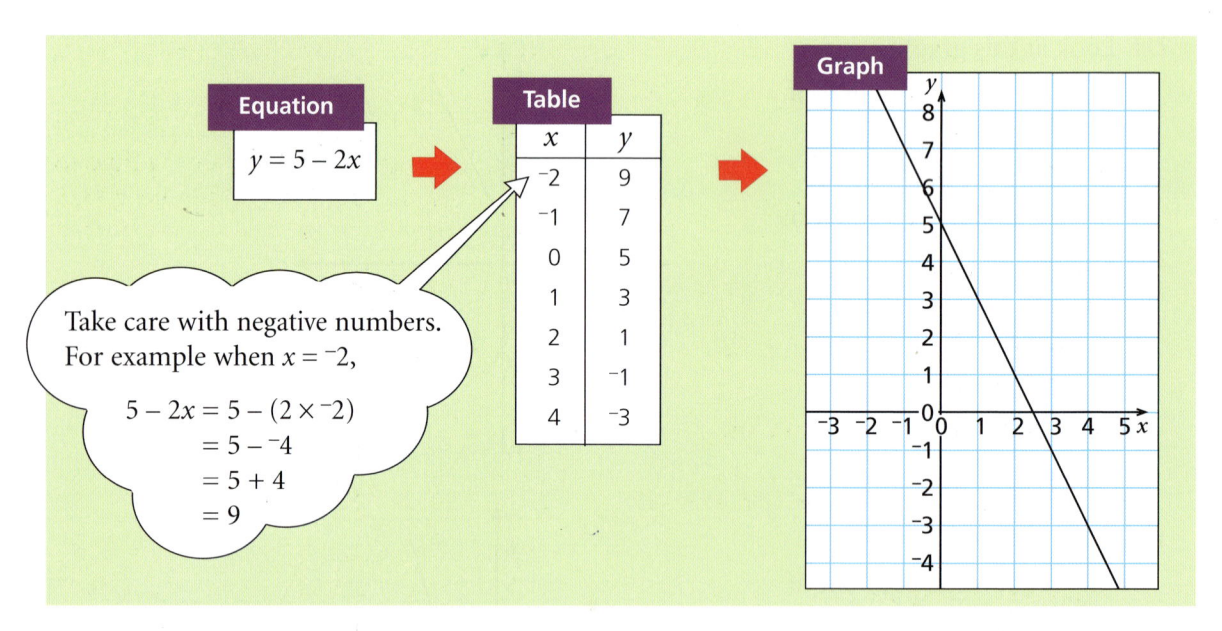

Equation

$y = 5 - 2x$

Table

x	y
-2	9
-1	7
0	5
1	3
2	1
3	-1
4	-3

Take care with negative numbers.
For example when $x = -2$,

$5 - 2x = 5 - (2 \times {}^-2)$
$= 5 - {}^-4$
$= 5 + 4$
$= 9$

Graph

E1 Draw the graph of $y = 3 - x$ like this.

(a) Copy and complete this table for $y = 3 - x$.

(b) Draw suitable axes for this graph.

(c) Plot the points you found in (a).
Check that they lie in a straight line.

(d) Draw and label the line.

x	y
-2	
0	
2	
4	

E2 For each of these equations,
- find some points that fit the equation
- draw suitable axes on squared paper
- plot the points; draw and label the graph

(a) $y = 4 - 2x$ (b) $y = 2 - 3x$
(c) $y = 6 - x$ (d) $y = {}^-4x + 7$

> Take care with negative numbers. For example when $x = {}^-2$,
> $${}^-4x + 7 = ({}^-4 \times {}^-2) + 7$$
> $$= 8 + 7$$
> $$= 15$$

> To measure the gradient of this line …

> … think of steps that go along **1 unit** each time …

> … and look at the height of each step.

The steps go **down** 3 units each time so we say the **gradient** of the line is **⁻3**.

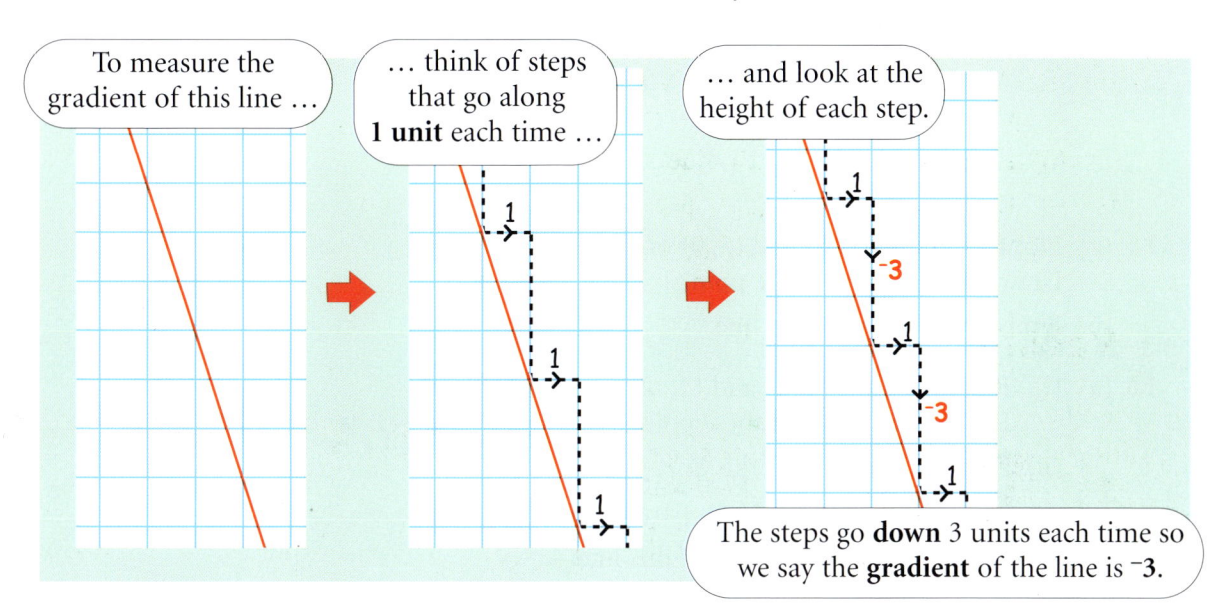

E3 Some lines with their equations are shown on the graph.

(a) What is the gradient of the line with equation $y = 8 - 2x$?

(b) Copy and complete this table.

Equation	Gradient	y-intercept
$y = 10 - x$		10
$y = 8 - 2x$		
$y = {}^-3x + 6$		
$y = {}^-1 - 4x$		

(c) What is the link between the equations of these lines and their gradients?

(d) What is the link between the equations of these lines and their y-intercepts?

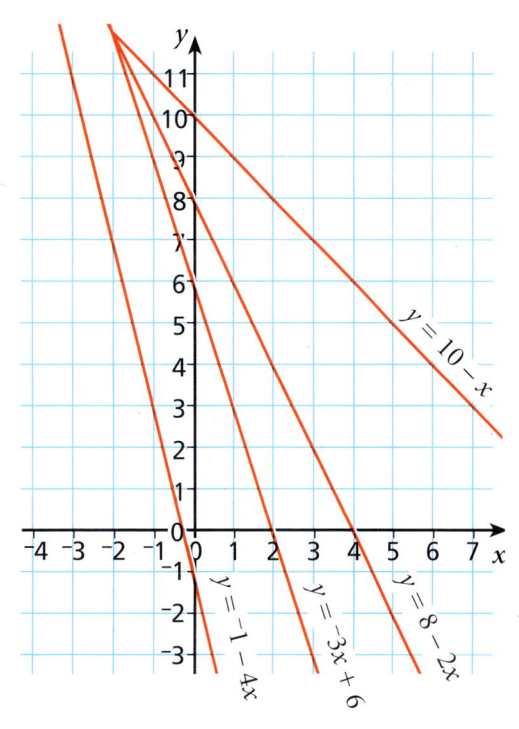

E4 (a) (i) What is the gradient of line A?
  (ii) What is its *y*-intercept?
  (iii) Write down its equation.
(b) Find the equation of line B.

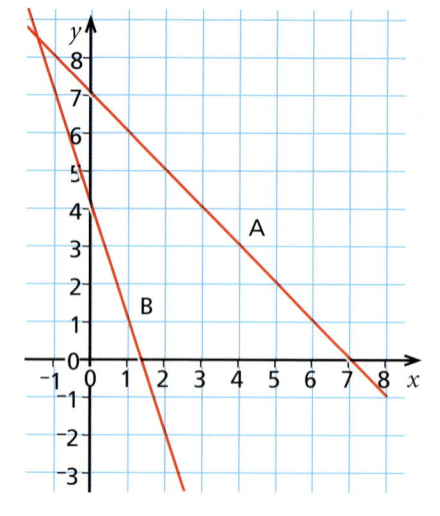

E5 (a) Plot the points (⁻1, 6) and (5, 0) on a pair of axes. Draw a line through these points.
(b) Find the equation of this line.

E6 (a) (i) Plot the points (1, 4) and (4, ⁻2) on a pair of axes. Draw a line through these points.
  (ii) Find the equation of this line.
(b) (i) Draw a line through (1, 0) parallel to the first line.
  (ii) Write down the equation of this line.

E7 Find the equation of each line drawn on the graph below.

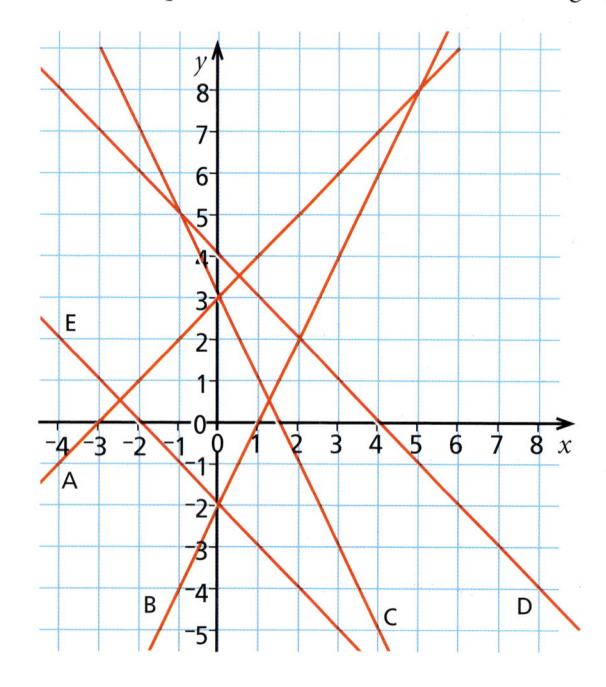

F Implicit equations

Can you draw graphs of these equations?

A $x + y = 10$ **B** $y - x = 6$ **C** $y + 2x = 5$ **D** $y - 3x = 4$

F1 Draw and label graphs of the following equations.
(a) $x + y = 8$ (b) $y - x = 6$ (c) $y - 2x = 3$ (d) $y + 3x = 9$

F2 (a) Copy and complete the table of values below for the equation $x + y = 3$.

x	−1	0	1	2	3	4
y						

(b) Make a table of values for the equation $y = 3 - x$.
(c) What do you notice about your tables?
Can you explain this?

F3 A straight line has equation $y - x = 5$.
(a) Rearrange the equation so that y is the subject (the equation should begin $y = \ldots$).
(b) What is the gradient of the line?
(c) Write down the y-intercept of the line.

F4 A straight line has equation $y + 2x = 5$.
(a) Rearrange the equation so that y is the subject.
(b) What is the gradient of the line?
(c) Write down the y-intercept of the line.

F5 Which two of these equations give a pair of parallel lines?

A $x + y = 5$ **B** $y = 2x + 4$ **C** $y - x = 6$ **D** $y - 2x = 7$

F6 On squared paper, draw a pair of axes both numbered from −2 to 8.
(a) On your axes, draw and label the line $y = x + 4$.
(b) On the same axes, draw and label the line $x + y = 2$.
(c) What are the coordinates of the point where the line $y = x + 4$ meets $x + y = 2$?
(d) On the same axes, draw the line $y = x - 2$.
What are the coordinates of the point where this line meets $x + y = 2$?
(e) If you draw one more line on your diagram, you can form a square.
Draw this line.
What is its equation?

What progress have you made?

Statement

Evidence

I can find the connection between tables and equations.

1 Which of these equations fits this table?

x	y
1	2
2	5
3	8
4	11

   A $y = 2x$ B $y = 3x - 1$
   C $x + y = 3$ D $y - x = 1$

I can draw the graph of an equation.

2 (a) For $y = 2x - 3$, what is y when
     (i) $x = 3$ (ii) $x = 2$ (iii) $x = 0$

   (b) On squared paper, draw axes numbered from ⁻5 to 5 and draw the graph of the line with equation $y = 2x - 3$.

I can use the link between the equation of a line, its gradient and y-intercept.

3 Which of the equations below gives
  (a) a line with a gradient of 2
  (b) a line with a y-intercept of 5
  (c) a line with a gradient of ⁻5

   A $y = 5x - 2$ B $y = 2x + 3$
   C $y = 3x + 5$ D $y = 3 - 5x$

I can work out the equation of a line.

4 (a) What is the gradient of this line?
  (b) What is the y-intercept of this line?
  (c) What is the equation of this line?

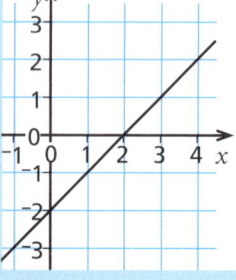

5 (a) (i) On a pair of axes, plot and join the points (3, 10) and (⁻1, ⁻2).
    (ii) Find the equation of this line.
  (b) (i) Draw a line through (0, ⁻4) parallel to the first line.
    (ii) Write down the equation of this line.

I can work with equations in different forms.

6 (a) On a pair of axes, draw the line with equation $y + 2x = 7$.

  (b) Which equations give lines parallel to this?
   A $y = ⁻2x + 1$ B $y = 2x + 3$
   C $y = 7 - 3x$ D $y = 8 - 2x$

11 Points, lines and arcs

This work will help you
- understand the idea of a locus (a set of points that follow some rule)
- make accurate drawings and get information from them

A Sets of points

An introductory activity is described in the teacher's guide.

Sheet 263 has a plan of a field, drawn to a scale of 1 cm to 1 metre.

A1 This message gives clues about where a sword is buried in the field.

The sword is 9 metres from corner A.
It is 6 metres from the fence BC.

(a) Use a pair of compasses to show all the possible places given by this clue.

(b) Use a ruler and set square to show all the possible places given by this clue.

(c) Put a cross where the sword is buried and label it 'sword'.

(d) How far in metres is the sword from corner B?

A2 *The helmet is 11 metres from corner D and 7 metres from corner C.*

Carefully do a construction to find where the helmet is buried.
Put a cross and label it 'helmet'. Do not rub out your construction.

A3 *The crown is 4 metres from fence AB and 5 metres from fence AD.*

Find the crown by drawing. Put a labelled cross where the crown is.

A4 *The chest of gold is less than 12 metres from A and less than 8 metres from D. To find the exact spot, you will have to do quite a lot of digging.*

Draw accurately the boundary of the part of the field you will need to dig.
Shade this area.

B Accurate drawing

B1 This is a sketch of a large, flat, hexagonal kite.

(a) Using a ruler, compasses and angle measurer, do an accurate drawing of the kite to a scale where 1 cm represents 20 cm.

(b) Measure the length of edge CD on your drawing. What will this be on the real kite?

(c) Measure the angle at vertex B. What will this be on the real kite?

(d) Describe any symmetry that the kite has.

B2 Two coastguard stations, P and Q, are 12 km apart on a straight coastline. P is exactly to the north of Q.

Coastguard P sees boat B on a bearing of 120°.

At the same moment, coastguard Q sees the boat on a bearing of 042°.

(a) Draw an accurate map of the situation to a scale where 1 cm represents 1 km.

(b) Measure these distances on the map and work out what they would be in real life.

   (i) The distance from the boat to coastguard P

   (ii) The distance from the boat to coastguard Q

   (iii) The distance from the boat to shore by the shortest route

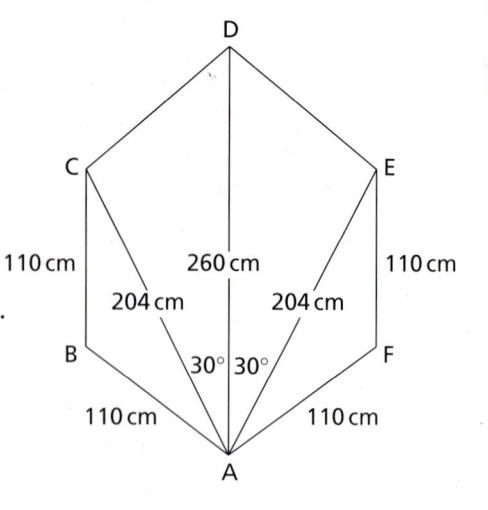

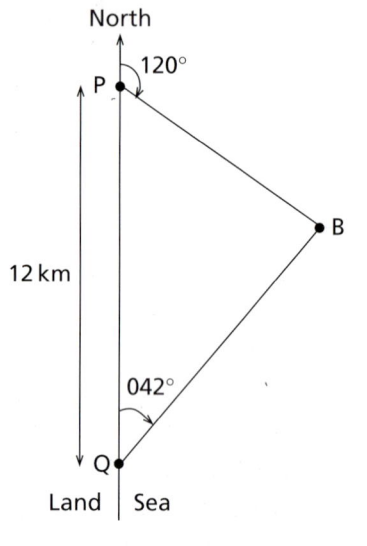

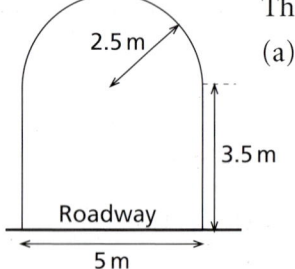

B3 This sketch shows an arch of a bridge going over a road.

(a) Draw the arch accurately to a scale where 2 cm represents 1 metre.

(b) Assume that any lorry going under the arch will be 3.0 m wide and will stay in the middle of the road. From your drawing, find the maximum height of lorry that will go under the arch without hitting it.

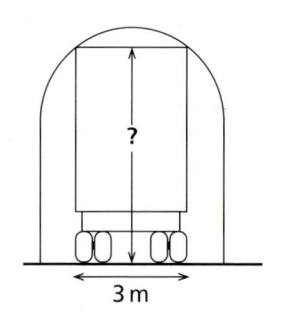

B4 A ladder is leaning against a wall.
The ground is horizontal.
The problem is to find how far the foot of the ladder is from the wall.

(a) Follow these instructions to make a scale drawing.
Use a scale where 2 cm represents 1 m.

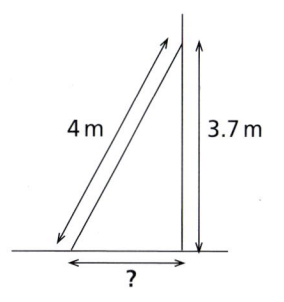

> Draw straight lines for the ground and wall, at right angles to each other.
>
> On the wall, mark the top of the ladder, the correct distance from the ground.

> Set your compasses to a radius that represents the length of the ladder.
>
> Put the compass point where you marked the top of the ladder.
>
> Construct an arc that cuts the ground.

> Draw in the ladder.

(b) Measure the distance of the foot of the ladder from the wall.

(c) What would it be full size?

B5 Carol is flying a kite on a horizontal field.
The kite string is unwound to its full length of 25 m.
David stands underneath the kite.
He is 18 m from Carol.

Use a scale drawing to find how high the kite is above the ground. (Assume Carol's hand holding the string is 1 m above the ground.)

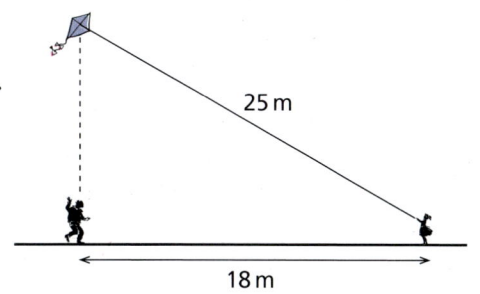

B6 A helicopter flies 85 km on a bearing of 208°.
Use a scale drawing to find these distances.

(a) How far south it has gone   (b) How far west it has gone

B7 These are the measurements of a quadrilateral shaped garden.

(a) Draw it accurately to a scale where 1 cm represents 1 metre.

(b) Take measurements to find the area of each triangular part.

(c) What is the area of the whole garden?

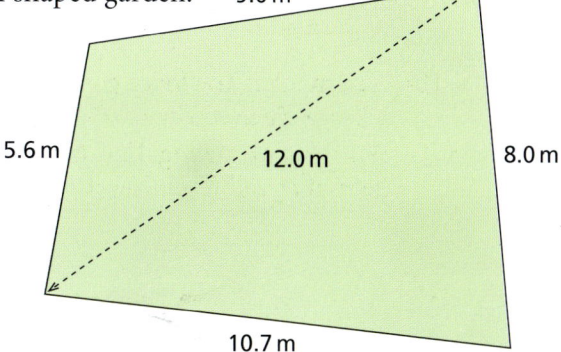

Your own scale drawing

Make your own drawing (plan or side view) of an object that interests you.
It must be something you can draw using just straight lines and arcs.

Start by carefully measuring lengths and angles on your object.
Choose a scale so your drawing fits on the page.

You could use a drawing program to do your drawing.

What progress have you made?

Statement

I can use the idea of a locus.

Evidence

1 These are the dimensions of a rectangular garden with high walls.

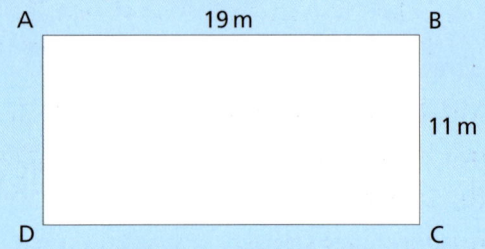

```
A        19 m        B
┌────────────────────┐
│                    │
│                    │ 11 m
│                    │
└────────────────────┘
D                    C
```

(a) Draw a plan of the garden to a scale where 1 cm represents 2 m.

(b) The gardener has a shrub that needs 'a sunny position' and decides to plant it at least 3 m from all the walls.
Show by drawing and shading where it can be planted.

(c) The gardener also wants the shrub to be 6 m from corner A.
Show clearly where it can be planted.

I can use scale drawing to solve problems.

2 This is a plan of a very small bedroom.

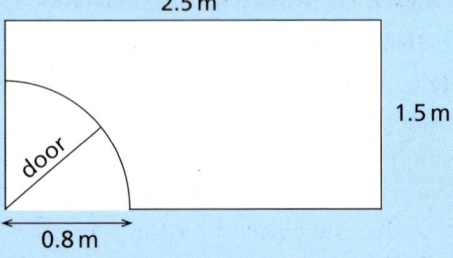

```
       2.5 m
  ┌─────────────┐
  │╲            │
  │ ╲           │ 1.5 m
  │door         │
  └─────────────┘
  ←0.8 m→
```

The occupant wants to put in a bed 2 m long.
What is the widest the bed can be for the door still to open fully?

12 Percentage problems

This work will help you
- express increases or decreases as a percentage
- choose the correct methods to solve a variety of percentage problems

A Review: increasing and decreasing

A1 (a) Which of these multipliers will **increase** an amount by 15%?

× 0.15   × 115   × 1.15   × 15   × 100.15

(b) Mr Fraser decides to increase his children's pocket money by 15%.
Before the increase, Alastair got £2.00 and Siobhan got £3.00 each week.
Calculate their pocket money after the increase.

A2 Match each percentage increase to its multiplier.

(a) Increase by 4%   A × 1.4   B × 1.14   C × 1.04
(b) Increase by 40%
(c) Increase by 14%

A3 Kay is a nurse who is paid £18 500 each year.
One year, her pay increases by 4%.
How much will she be paid each year after the increase?

A4 (a) Which of these multipliers will **decrease** an amount by 28%?

× 0.28   × 1.28   × 0.72   × 1.72   × 0.82

(b) The value of a car decreased by 28% over the last year.
The car was worth £4500 a year ago.
How much is the car worth now?

A5 Match each percentage decrease to its multiplier.

(a) Decrease by 63%   A × 0.7   B × 0.37   C × 0.97
(b) Decrease by 30%
(c) Decrease by 3%

A6 A shop reduces its prices by 30% in a January sale.
What is the sale price of a frying pan that used to cost £40?

A7 (a) Increase 50 kg by 42%.   (b) Decrease 50 kg by 42%.

B Finding an increase as a percentage

These diagrams show three young people at ages 10 and 16 years.

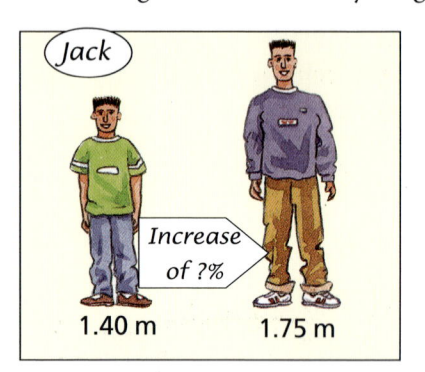

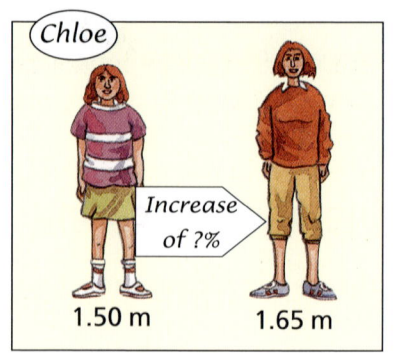

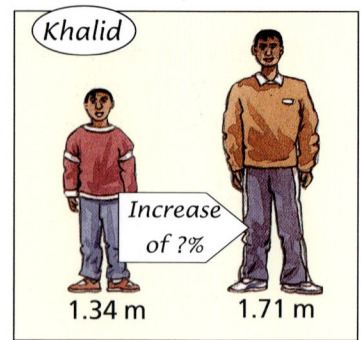

Jack	Chloe	Khalid
1.40 m → 1.75 m Increase of ?%	1.50 m → 1.65 m Increase of ?%	1.34 m → 1.71 m Increase of ?%

Give all answers correct to the nearest 1%.

B1 When Bindoo started secondary school her height was 1.4 metres.
Her height was 1.61 metres when she left.
What was the percentage increase in her height during this time?

B2 In Scott's first job, he started on wages of £250 per week.
After a year his wages rose to £270 per week.
What was the percentage increase in Scott's wages?

B3 A shop increases the price of a camera from £150 to £180.
What was the percentage increase in the price of the camera?

B4 These weight increases are for some animals in one month early on in their lives.

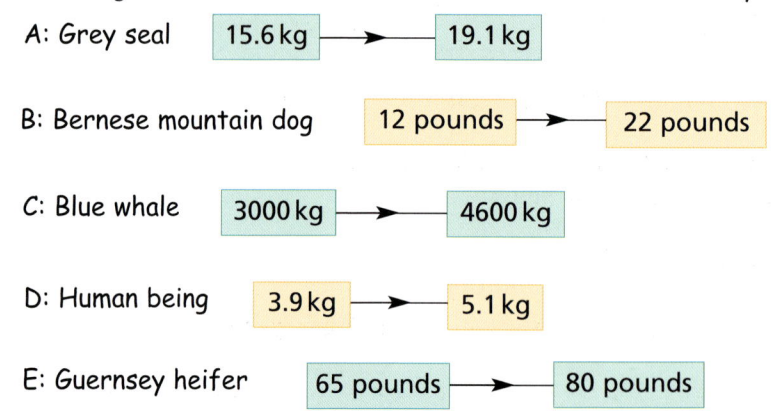

A: Grey seal 15.6 kg → 19.1 kg

B: Bernese mountain dog 12 pounds → 22 pounds

C: Blue whale 3000 kg → 4600 kg

D: Human being 3.9 kg → 5.1 kg

E: Guernsey heifer 65 pounds → 80 pounds

(a) Without using a calculator, which animal do you think shows the largest percentage increase in weight?

(b) Calculate each percentage increase in weight.
Comment on your answer to (a).

B5 In 1990 our average yearly consumption of chicken and turkey was 20.9 kg per person. By 1999 it had risen to 27.7 kg. What was the percentage increase in our chicken and turkey consumption?

B6 Sue sometimes travels from Taunton to London for business meetings.
In 1997 her return rail fare cost £73.40.
In 2002 the same journey cost £96.20.
What was the percentage increase in Sue's rail fare between those years?

B7 In 1999 about 269 000 people booked a package holiday to Italy.
In 2000 this figure rose to 285 000 people.
What was the percentage increase in these bookings between 1999 and 2000?

B8 When Ranjit was ten years old he weighed 32.9 kg.
When he was only five he weighed 19.3 kg.
What was the percentage increase in his weight during this time?

C Finding a decrease as a percentage

Emma buys a car for £5600.
A year later the car is worth £4760.

- What is the multiplier here? £5600 → × ? → £4760

- What is the percentage decrease in the value of Emma's car?

Give all answers to the nearest 1%.

C1 (a) Work out the multiplier for this diagram. £250 → × ? → £185

(b) Write down the percentage decrease from £250 to £185.

C2 In a sale the price of a book is cut from £5.00 to £3.30.
Calculate the percentage decrease in the price.

C3 Brian buys a computer for £1400.
A year later it is worth only £840.
What is the percentage decrease in the value of Brian's computer?

C4 (a) Work out the multiplier for the diagram below, correct to 2 decimal places.

14 kg → × ? → 5 kg

(b) What is the percentage decrease from 14 kg to 5 kg?

C5 In a sale, a shop reduces some of its prices.

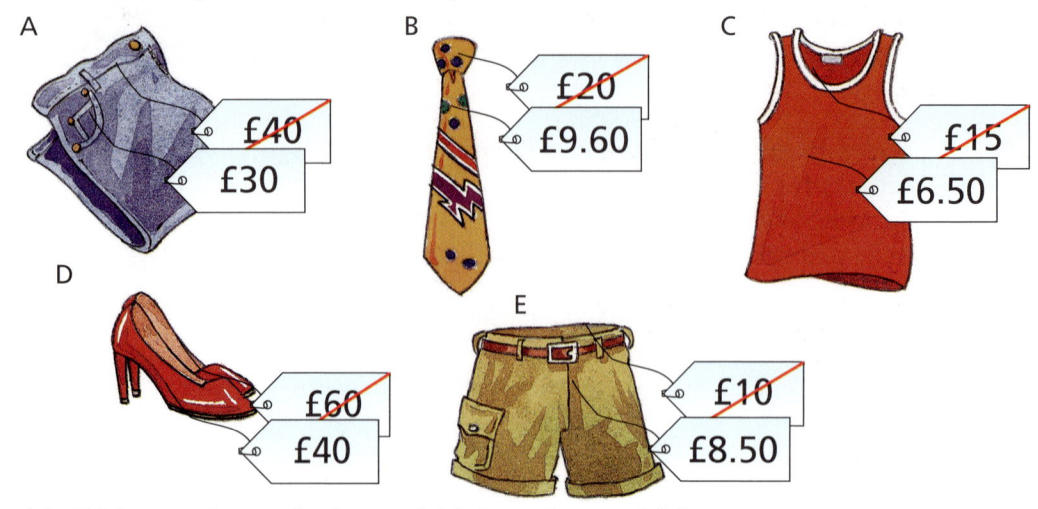

A £40 £30

B £20 £9.60

C £15 £6.50

D £60 £40

E £10 £8.50

(a) Without using a calculator, which item do you think is reduced by the largest percentage?

(b) Calculate each percentage reduction in price.
Comment on your answer to (a).

C6 In 1998 there were 3144 convictions for cruelty to animals.
By 1999 the number of convictions had fallen to 2719.
Find the percentage decrease in convictions between 1998 and 1999.

C7 In 1999 about 470 000 people booked a package holiday to Portugal.
In 2000 this figure fell to 440 000 people.
What was the percentage decrease in these bookings between 1999 and 2000?

D Mixed increases and decreases

Give all answers to the nearest 1%.

D1 In 1982 there were 1033 kidney transplants in the UK.
By 1995 the number of kidney transplants had risen to 1796.
Find the percentage increase in the number of these transplants between 1982 and 1995.

D2 Between 1983 and 1994 the number of cinema visits in Britain rose from 66 million to 124 million.
Find the percentage increase in the number of cinema visits between 1983 and 1994.

D3 In 1975 the number of cases of scarlet fever in the UK was 10 235.
In 1955 the number of cases of scarlet fever was 38 850.
Work out the percentage decrease in the number of cases during this period.

D4 (a) Between 1994 and 1995 the number of domestic burglaries in England and Wales fell from 680 735 to 646 700.
Calculate the percentage decrease in the number of domestic burglaries in England and Wales between 1994 and 1995.

(b) Over the same period in Scotland, the number of domestic burglaries fell from 53 244 to 44 725. Calculate the percentage decrease in the number of domestic burglaries in Scotland.

D5 (a) In 1985 the average distance a person walked each year was 244 miles. In 1999 this figure was 191 miles.
Find the percentage decrease in the average distance walked between 1985 and 1999.

(b) Over the same period the average distance a person travelled each year in total rose from 5317 miles to 6806 miles.
Find the percentage increase in the average distance travelled each year.

(c) Comment on your results.

D6 The number of vehicles stolen in 1991 was 634.
In 1999 the number of vehicles stolen was 429.
Did the number of vehicles stolen increase or decrease during this period?
What was the percentage change in the number of stolen vehicles?

D7 (a) In 1985 there were 559 200 pupils in private schools in Britain.
In 1995 there were 586 300 such pupils.
Calculate the percentage change in the numbers of private school pupils.
Was the percentage change an increase or decrease?

(b) Over the same period the total number of pupils at school in the UK rose from 9 565 000 to 9 707 000.
Calculate the percentage change in the numbers of school pupils.

(c) Comment on your results.

E Mixed problems

Give all answers to the nearest 1%.

E1 A bus company plans to increase its fares by 5%.
Calculate the new price of a ticket that cost £1.20 before the increase.

E2 Meena scored 21 out of 28 in a French test. What was her score as a percentage?

E3 The percentage of sugar in Joshua's breakfast cereal is 17%.
How much sugar is in a 50 g serving of his cereal?

E4 Between 1990 and 2000 the population of Leeds rose from 712 800 to 726 100.

(a) To the nearest 1%, find the percentage increase over these ten years.

(b) If the population increases by this percentage over the next ten years, what will be the population of Leeds in 2010, to the nearest thousand?

E5 In a 1999 survey, 18 294 pupils were asked, 'What would you do to make schools better?' 4663 pupils replied, 'More resources and better facilities'.
What percentage of pupils in the survey gave this reply?

*E6 This table gives information about live births in England and Wales by age of mother at birth.

Age group of mother	Number of live births	
	1981	1999
0–19	56 600	48 400
20–24	194 500	110 700
25–29	215 800	181 900
30–34	126 600	185 300
35–39	34 200	81 300
40+	6900	14 300
Total	634 600	621 900

(a) What percentage of babies born in 1981 were to mothers in the 20–24 age group?
(b) What percentage of babies born in 1999 were to mothers that were younger than 20?
(c) Find the percentage decrease in the total number of births between 1981 and 1999.
(d) For the following age groups, find the percentage change in the number of babies born between 1981 and 1999. State if each change was an increase or a decrease.
 (i) 25–29 (ii) 30–34 (iii) 40+

What progress have you made?

Statement

I can express an increase as a percentage increase.

I can express a decrease as a percentage decrease.

I can solve percentage problems.

Evidence

1 A shop increases the price of a kettle from £24.50 to £29.40.
What was the percentage increase on the price?

2 When Salma was born she weighed 2.95 kg. After a few days she weighed 2.66 kg.
What was the percentage decrease in her weight?

3 In 2000 the population of South Africa was 20.6 million.
About 4.1 million of them were HIV positive.
What percentage of the population is this?

4 An unstretched metal spring is 15 cm long.
It is stretched to a length of 19 cm.
Find the percentage change in its length.

Review 2

1. A cake has a diameter of 30 cm.
   Mark has a piece of ribbon 95 cm long.

   Is the ribbon long enough to fit round the cake?
   Show all your working clearly.

2. In the diagram triangle Y is an enlargement of triangle X.
   (a) What is the scale factor of this enlargement?
   (b) Which point, P, Q, R or S, is the centre of enlargement?

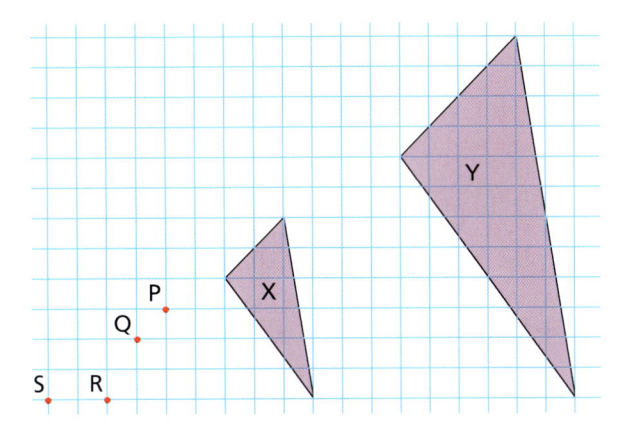

3. Fiona is paid £25 000 each year.
   One year her pay increases by 5%.

   How much will she be paid after the increase?

4. (a) (i) Which of the lines, A or B, has a gradient of 5?
       (ii) Write down the equation of this line.
   (b) What is the equation of the other line?

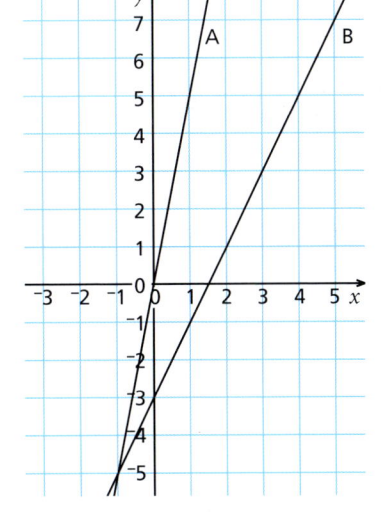

5. Use the clues to find the value of the whole number n each time.

   (a)
   - n is prime
   - $n + 1$ is odd

   (b)
   - $\frac{6n}{2}$ is even
   - n is less than 10
   - n is a square number

   (c)
   - $\frac{n}{5} + \frac{1}{5}$ is a whole number
   - n is prime
   - n is less than 20

6 Joe sometimes travels on the London Underground.
One year the cost of the cheapest ticket rose from £1.50 to £1.70.
Work out the percentage increase in the price of this ticket, correct to the nearest 1%.

7 Kenny's garden is a rectangle 12 m by 9 m.
The point T shows the position of a tree.

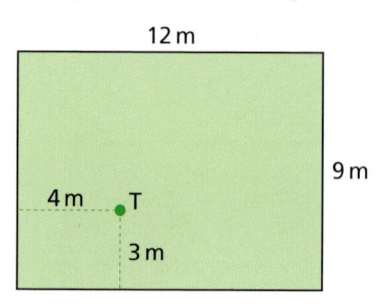

He wants to plant another tree in the garden.
It must be at least 2 m from the edges of the garden.
It must also be at least 4 m from the other tree.

Using a scale of 1 cm to 1 m draw a diagram of Kenny's garden and shade the region where the new tree could be planted.

8 Stephen buys a car for £15 000.
Two years later it is valued at £10 500.
Work out the percentage decrease in the value of his car.

9 What is the radius of a circle that has a circumference of 20 cm?
Give your answer to 1 d.p.

10 (a) Draw the graph of the line with equation $y + 2x = 8$.
(b) What is the gradient of this line?

11 Sue wants to make a set of bookshelves to fit in the space under her stairs.
The sketch shows the side view of the space.

The bookshelves are to be 1.32 m high.

Use a scale drawing to decide on the maximum possible width for Sue's bookshelves.
Use a scale of 1 cm to 20 cm.

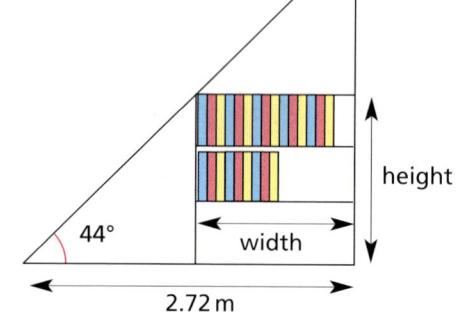

12 During 1999 there were 569 people drowned in the UK.
A total of 248 people drowned in rivers or streams.
Out of the people who drowned, what percentage of them drowned in a river or stream?

13 **A** $y + x = 8$ **B** $y + 3x = 10$ **C** $y - 3x = 6$ **D** $y - 5 = 3x$

(a) Rearrange each of the above equations so that y is the subject.
(b) Which two of these equations give a pair of parallel lines?

⑬ Ratio and proportion

This work will help you
- write a ratio as a single number
- use ratios to decide if two quantities are in direct proportion
- solve problems involving direct proportion using algebra

A Single number ratios

This rectangular window is 50 cm high and 20 cm wide.
The ratio **height : width** is 50 : 20 or 5 : 2.

Another way to write the ratio is as a division: $\dfrac{\text{height}}{\text{width}}$

So, for this window, the ratio $\dfrac{\text{height}}{\text{width}} = \dfrac{50}{20} =$ **2.5**

Written this way, the ratio is a **single number**.
The ratio 2.5 tells you that the height is 2.5 times the width.

A1 Calculate the ratio $\dfrac{\text{height}}{\text{width}}$ for each of these windows.

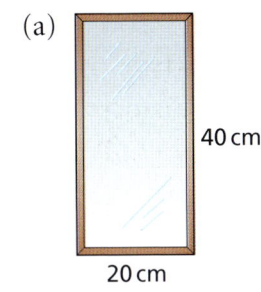

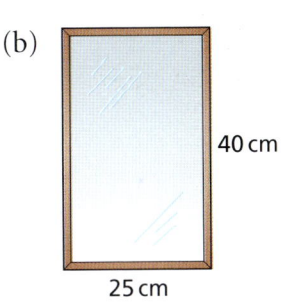

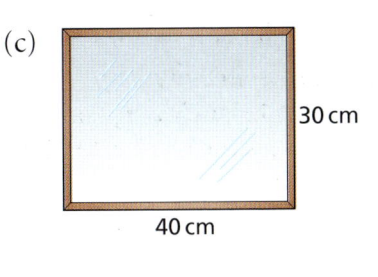

(a) 40 cm high, 20 cm wide
(b) 40 cm high, 25 cm wide
(c) 30 cm high, 40 cm wide

A2 A window has the ratio $\dfrac{\text{height}}{\text{width}} = 1$. What shape is the window?

A3 For each animal below, calculate the ratio $\dfrac{\text{weight of daily food intake}}{\text{body weight}}$.

Animal	Body weight	Weight of daily food intake
Hamster	125 g	25 g
Cat	3000 g	300 g
Elephant	5000 kg	20 kg

A4 An average 18-year-old male is 175 cm tall.
An average 1-year-old male is 74 cm tall.

Calculate the ratio $\dfrac{\text{height of 18-year-old male}}{\text{height of 1-year-old male}}$ correct to one decimal place.

A5 A fully grown adult Nile crocodile is 500 cm long.
A newly born Nile crocodile is 26 cm long.

Calculate the ratio $\dfrac{\text{length of adult Nile crocodile}}{\text{length of newly born Nile crocodile}}$ correct to one decimal place.

A6 Here are some recipes for pink paint.

Fiesta pink	Passion pink	Hot pink	Baby pink
Mix red and white in the ratio 14 : 11	Mix red and white in the ratio 7 : 10	Mix red and white in the ratio 10 : 7	Mix red and white in the ratio 1 : 4

(a) For each recipe, work out the ratio $\dfrac{\text{amount of red paint}}{\text{amount of white paint}}$ correct to 2 d.p.

(b) Which recipe gives the deepest shade of pink?
How did you decide?

B Ratio and direct proportion

a	1	2	3	4
b	60	120	180	240

x	1	2	3	4
y	30	90	150	210

- Work out the ratio $\dfrac{b}{a}$ for each pair of values.
- What does this tell you?

- Work out the ratio $\dfrac{y}{x}$ for each pair of values.
- What does this tell you?

B1 The times (T hours) to roast ribs of beef of various weights (W kg) are shown below.

W	1	2	3	4
T	1.5	2.5	3.5	4.5

(a) Work out the ratio $\dfrac{T}{W}$ for each pair of values.
(b) Explain how your ratios show that T is not directly proportional to W.

B2 Geri found the volumes (V cm³) of some wooden blocks.
She then weighed each block to find its weight (W grams).
Here are her results.

V	20	30	35	45	50
W	16	24	28	36	40

(a) Work out the ratio $\frac{W}{V}$ for each pair of values.
(b) Explain how your ratios show that W is directly proportional to V.
(c) Find an equation connecting W and V that begins $W = \ldots$
(d) Use your equation to find the weight of a wooden block with a volume of 72 cm³.

B3 This table shows the number of copies (n) produced by a printer when run for certain lengths of time (t minutes).

t	2	6	8	10
n	7	21	28	35

(a) Work out the ratio $\frac{n}{t}$ for each pair of values.
(b) Explain how your ratios show that n is directly proportional to t.
(c) Find an equation connecting n and t.
(d) Use your equation to find the number of copies that would be run off in 34 minutes.

B4 For each table below,
- work out the ratio $\frac{b}{a}$ for each pair of values.
- decide if b is directly proportional to a.
  If b is directly proportional to a, write down an equation connecting b and a.

(a)

a	0.5	2.6	4.8	5.1
b	1.0	5.2	9.6	10.2

(b)

a	2	4	6	8
b	6	12	26	48

(c)

a	5	10	20	25
b	9	18	36	45

(d)

a	1	3	5	7
b	5	11	25	47

B5 For the values in each table, y is directly proportional to x.
Find the missing value in each table.

(a)
x	1.4	2.8	5.6
y	3.5	■	14.0

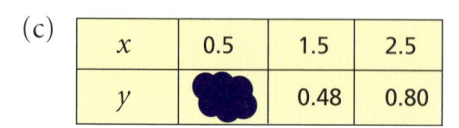

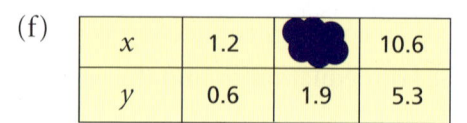

(b)
x	3.2	5.3	6.0
y	19.2	31.8	■

(c)
x	0.5	1.5	2.5
y	■	0.48	0.80

(d)
x	1.9	■	4.1
y	5.7	6.9	12.3

(e)
x	2.0	3.5	■
y	5.6	9.8	18.2

(f)
x	1.2	■	10.6
y	0.6	1.9	5.3

B6 (a) Explain how you can tell from the graph that p is **not** directly proportional to q.

(b) Confirm this by calculating the ratio $\frac{p}{q}$ for some pairs of values.

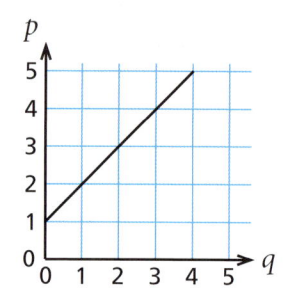

C Using algebra

You can sometimes solve a problem involving direct proportion by using ratios to form an equation and then solving it.

Example 1

Given that b is directly proportional to a, what is the value of k?

a	7	11
b	11.2	k

b is directly proportional to a

so $\quad \frac{k}{11} = \frac{11.2}{7}$

and $\quad k = \frac{11.2}{7} \times 11$

$\quad\quad\quad = 17.6$

Example 2

Tropical orange paint is made by mixing 11 parts of red paint with 6 parts of yellow paint.

How many litres of red paint would be needed to mix with 63 litres of yellow paint?

Let r be the number of litres of red paint

so $\quad \frac{r}{63} = \frac{11}{6}$

and $\quad r = \frac{11}{6} \times 63$

$\quad\quad\quad = 115.5$

so 115.5 litres of red paint are needed.

C1 For the table below, *y* is directly proportional to *x*.

x	1.2	8.6
y	7.8	

(a) Let the hidden value in the table be *n*.
    Use ratios to form an equation for *n*.

(b) Solve the equation to find the value of *n*.

C2 Photo B is an enlargement of photo A.

A: 7.5 cm (height), 10.5 cm (width)

B: k cm (height), 18.9 cm (width)

(a) Use ratios to form an equation for *k*.

(b) Solve the equation to find the value of *k*.

C3 Cool blue paint is made by mixing 12 parts of white paint with 5 parts of blue paint.
How many litres of white paint would be needed to mix with 42 litres of blue paint?

C4 Fiona is a jeweller.
She makes sterling silver for a ring by mixing 1.2 grams of copper with 14.8 grams of pure silver.

To make sterling silver for a necklace, how many grams of copper would she need to mix with 129.5 grams of pure silver?

C5 The graph below is a straight line going through (0, 0).
Find the value of *p*.

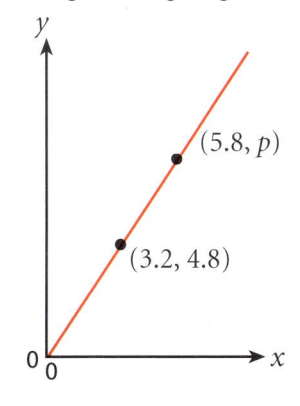

Points: $(3.2, 4.8)$ and $(5.8, p)$

What progress have you made?

Statement

I can calculate a ratio as a single number.

Evidence

1 Calculate the ratio $\frac{\text{height}}{\text{width}}$ for each photo, correct to 2 d.p.

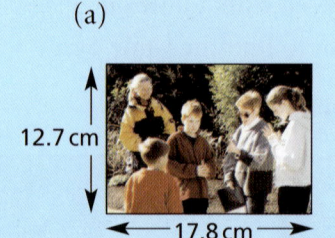

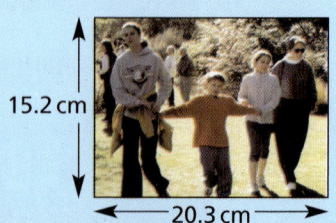

   (a) 12.7 cm, 17.8 cm

   (b) 15.2 cm, 20.3 cm

I can use the link between ratio and direct proportion.

2 The costs (£C) of some journeys of various distances (D km) on Southspeed Trains are shown below.

D	15	19	28	40
C	3.50	3.90	5.10	6.90

   (a) Find the ratio $\frac{C}{D}$ for each pair of values.

   (b) Explain how your ratios show that C is **not** directly proportional to D.

3 (a) Work out the ratio $\frac{y}{x}$ for each pair of values in the table below.

x	1.6	3.0	4.8
y	2.4	4.5	7.2

   (b) Explain how your ratios show that y is directly proportional to x.

   (c) Find an equation connecting y and x.

   (d) Use your equation to find y when $x = 3.4$.

I can use algebra to solve problems involving direct proportion.

4 In the table below, b is directly proportional to a.

a	2.5	9.5
b	8.5	n

   (a) Use ratios to form an equation for n.

   (b) Solve the equation to find the value of n.

14 Angles of a polygon

This is about the angles in shapes with several straight sides (polygons). The work will help you

- relate the angles of a polygon to the number of sides it has
- work out and use the angles of regular polygons

A Interior angles

- What is this type of polygon called?
- How many triangles has it been split into?
- What happens if you split it into triangles another way? (**The vertices of the triangles must be vertices of the polygon.**)

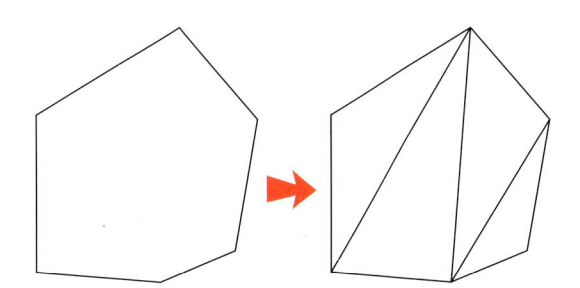

Look at each polygon below.

- How many sides does it have?
- What is its special name?
- How many triangles can you split it into?
- What is the sum of the interior angles?

What is the rule connecting the number of sides and the sum of the interior angles?

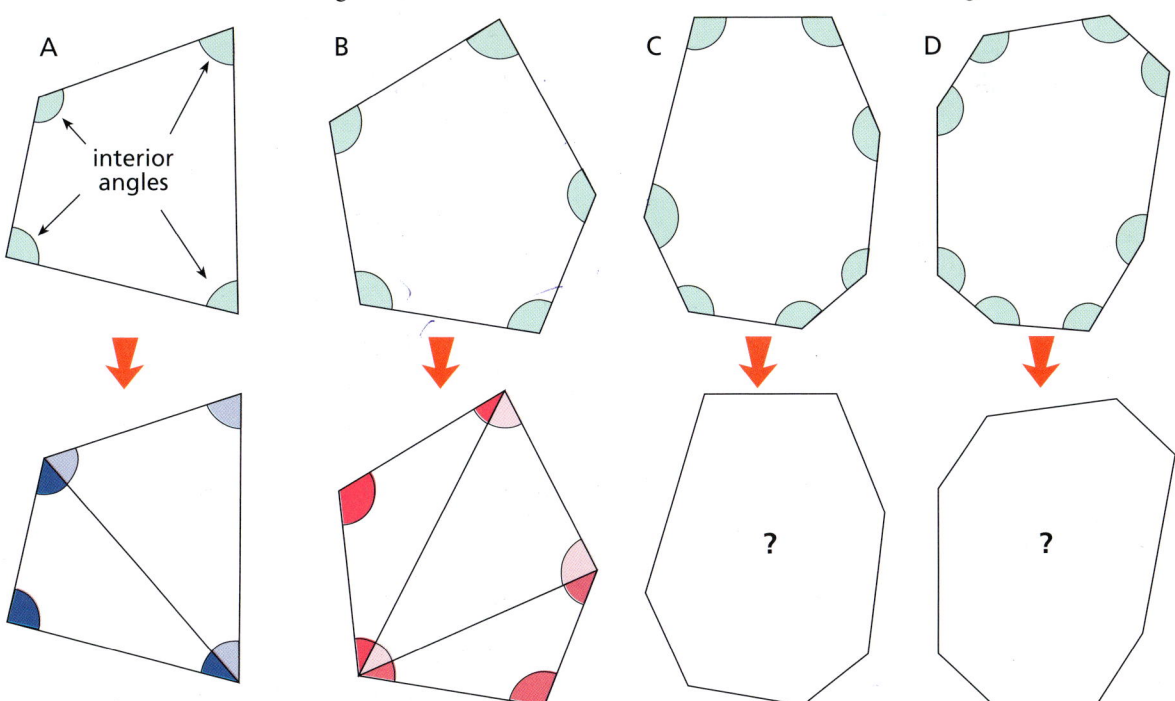

A interior angles

B

C ?

D ?

A1 (a) What should the sum of the interior angles of a nine-sided polygon be?

(b) Draw any nine-sided polygon (make it fairly large). Measure the interior angles and add them up. See how close the total is to your answer (a).

A2 Do the same as in A1 with a ten-sided polygon.

A3 For each of these polygons,
(a) count the number of sides
(b) say what the sum of the interior angles will be
(c) work out the missing angle

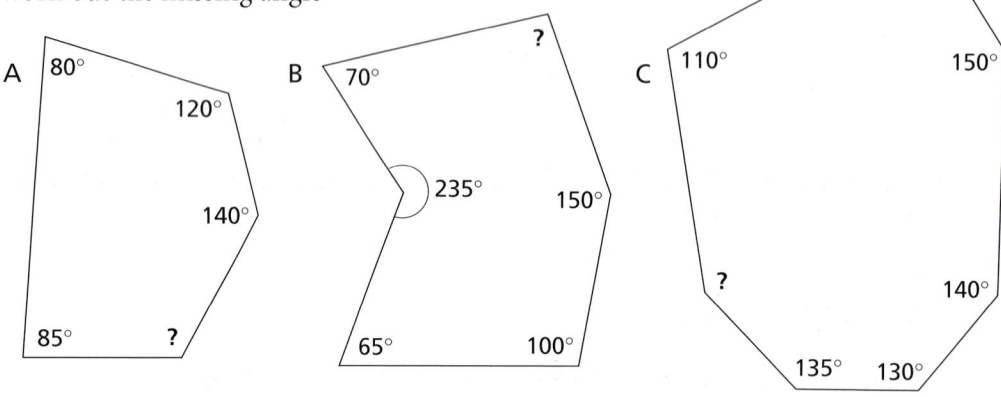

A: 80°, 120°, 140°, ?, 85°

B: 70°, ? , 235°, 150°, 100°, 65°

C: 150°, 130°, 150°, 140°, 130°, 135°, ?, 110°

A4 Work out the missing angles on this kite.

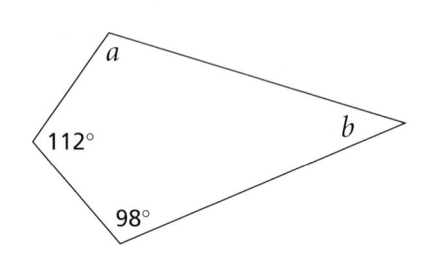

a, b, 98°, 112°

A5 Work out the angles that the letters stand for. (Angles with the same letter are the same size.)

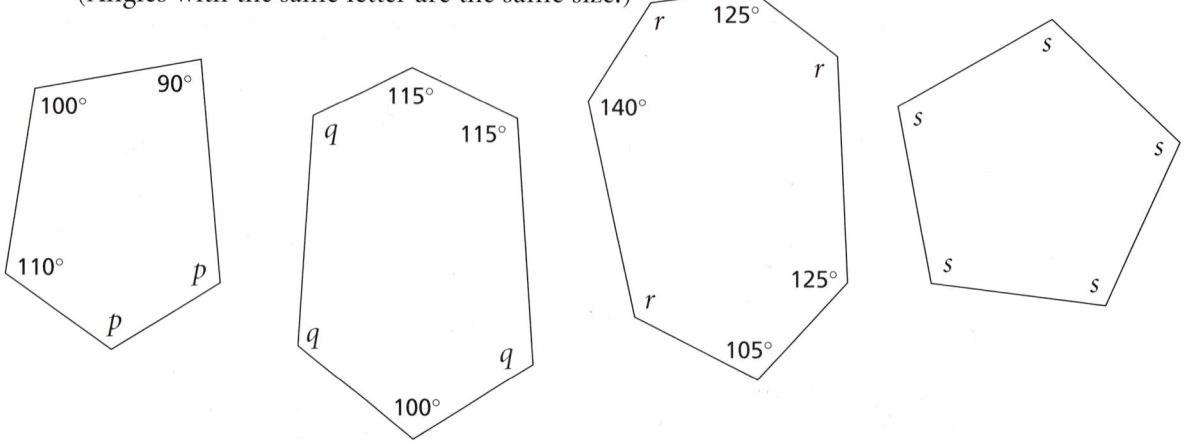

100°, 90°, 110°, p, p

115°, 115°, q, q, 100°, q

125°, 140°, 105°, 125°, r, r, r

s, s, s, s, s

A6 This hexagon has rotation symmetry of order 2.
Work out the angles marked with letters.

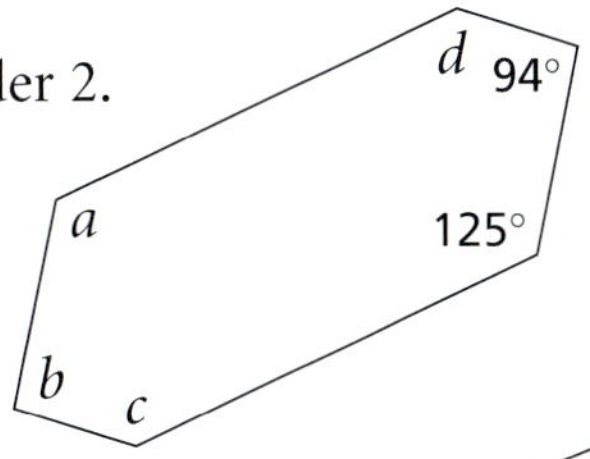

A7 Work out the missing angles here.

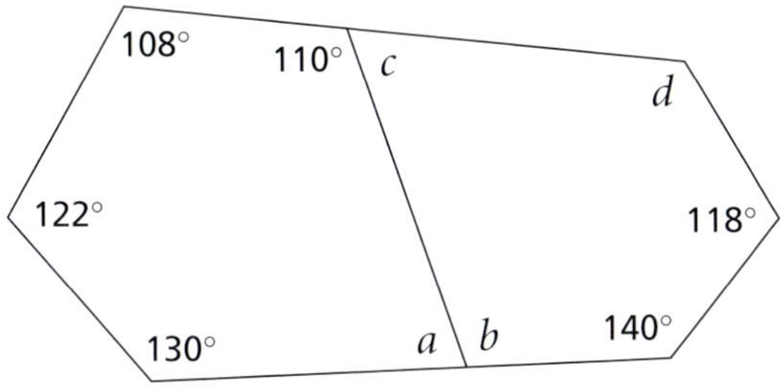

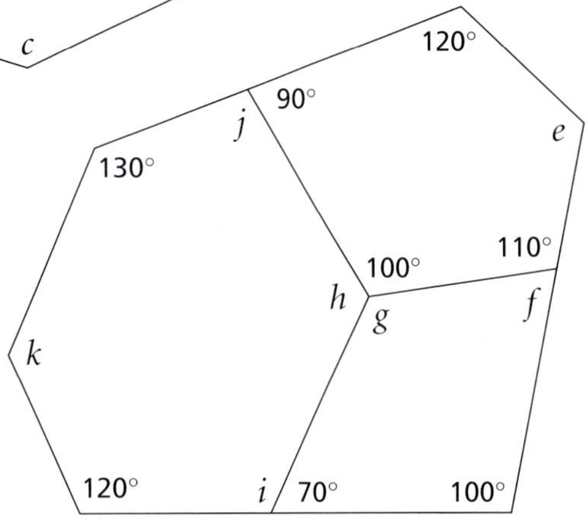

A8 Nadim measures the interior angles of a polygon and finds they add up to 2340°.
How many sides does the polygon have?

B Exterior angles

If you extend a side of a polygon, this angle is called an **exterior angle**.

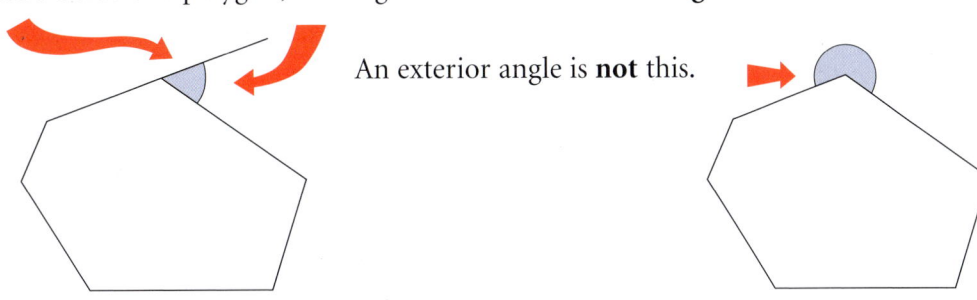

An exterior angle is **not** this.

B1 Work out the exterior angles marked with letters here.

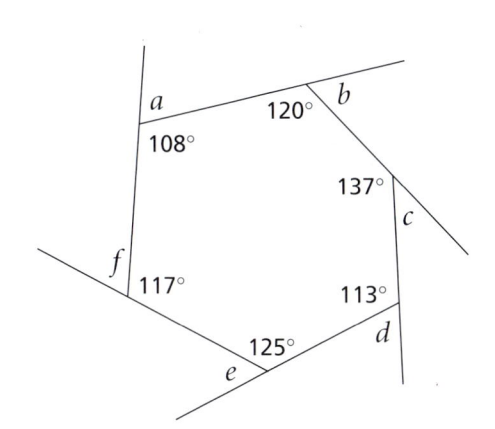

B2 Use the exterior angles to work out the interior angles marked with letters.

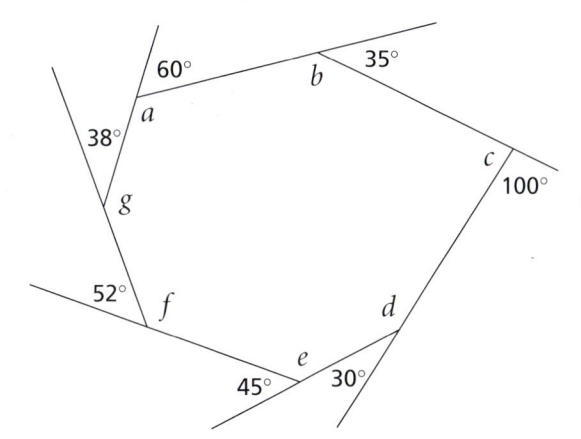

Exterior angles experiment

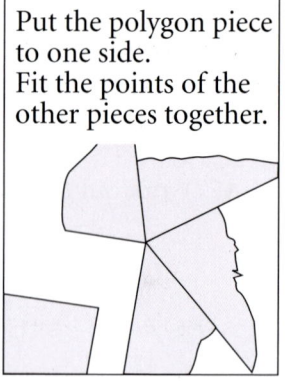

| Draw a polygon in the centre of a rough circle of paper. | Extend each side clockwise out towards the edge of the paper. | Cut along the drawn lines. | Put the polygon piece to one side. Fit the points of the other pieces together. |

- What happens?
- Why does it happen?
- Will this work for a polygon with any number of sides?

B3 Find the missing angles.

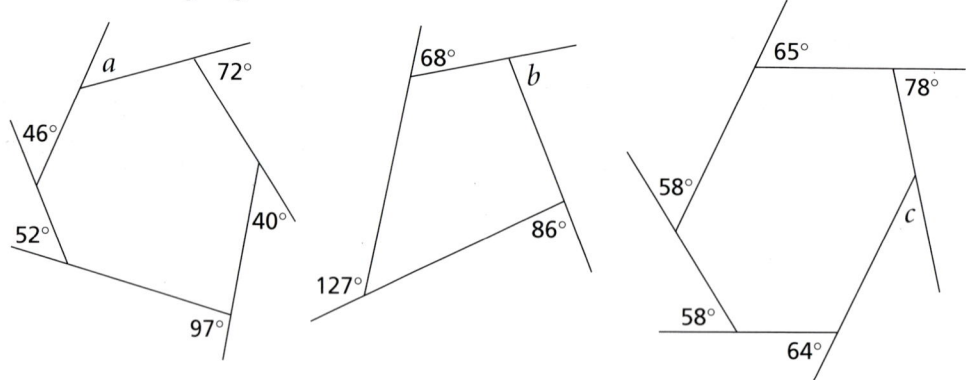

B4 Find the angles the letters stand for.
(Angles with the same letter are the same size.)

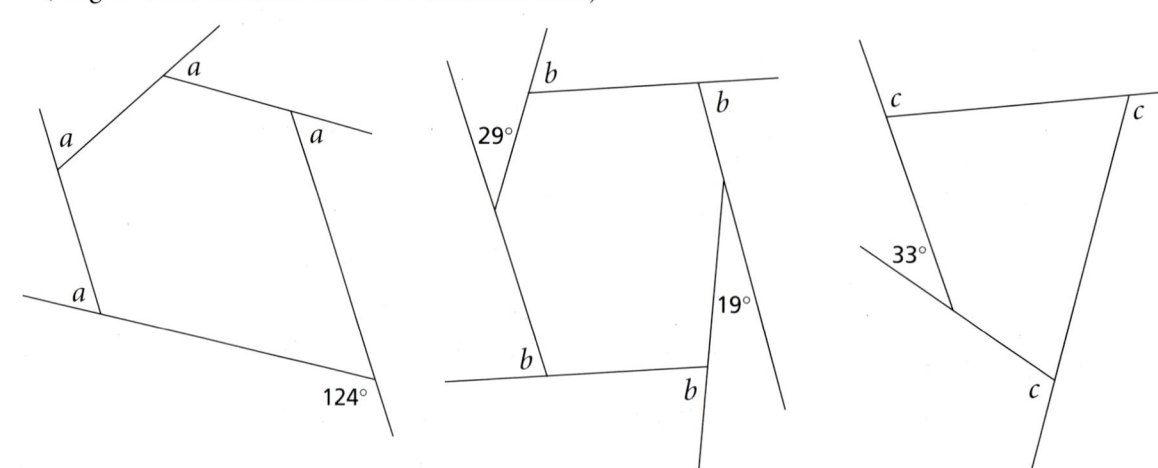

C Regular polygons

A **regular** polygon is one with all its sides the same length and all its interior angles equal.

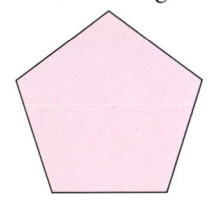

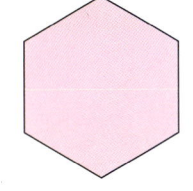

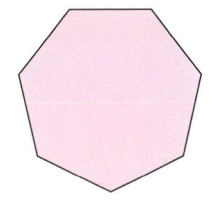

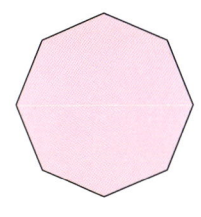

Regular pentagon   Regular hexagon   Regular heptagon   Regular octagon

C1 For a regular hexagon give
   (a) the sum of the **exterior** angles
   (b) the size of one exterior angle

C2 For a regular hexagon work out
   (a) the sum of the **interior** angles
   (b) the size of one interior angle

C3 Copy this table.

(a) Name of polygon	Number of sides	(b) Size of each exterior angle	(c) Size of each interior angle
	4		
	5		
	6		
	7		
	8		

   (a) Fill in column (a).
   (b) Put your answer to C1(b) in the right place in column (b).
       Work out the exterior angle for each other regular polygon and put the results in column (b) too.
   (c) Put your answer to C2(b) in the right place in column (c).
       Work out the interior angle for each other regular polygon and put the results in column (c) too.
   (d) What do you notice about the exterior and interior angles of each regular pentagon? Is this this what you expect?

C4 Work out the exterior and interior angles of
   (a) a regular nonagon (9 sides)   (b) a regular icosagon (20 sides)

C5 These are parts of regular polygons. How many sides does each one have?
   (a) 20°   (b) 24°   (c) 150°

C6 Each interior angle of a regular polygon is 144°.

  (a) What is each exterior angle?

  (b) How many sides does this regular polygon have?

C7 Work out how many sides a regular polygon has if each interior angle is

  (a) 165°   (b) 171°

C8

Draw a circle at least as big as an angle measurer.	Put the centre of an angle measurer on the centre of the circle. Mark off every 30°.	Draw 'spokes' through your marks.	Join the ends of the spokes to make a regular polygon.

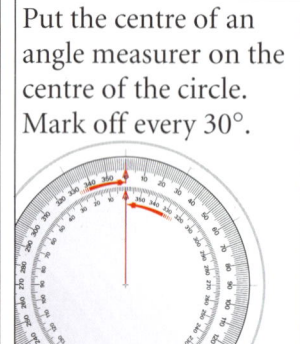

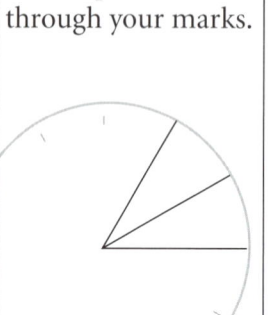

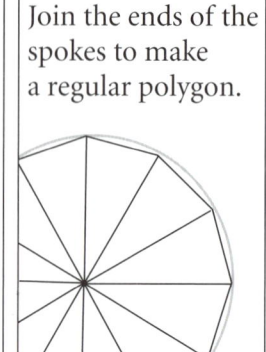

  (a) How many sides does your polygon have?

  (b) Extend the sides clockwise so exterior angles are showing.

  Mark all the angles equal to this one with a •.

  Mark all the angles equal to this one with a ×.

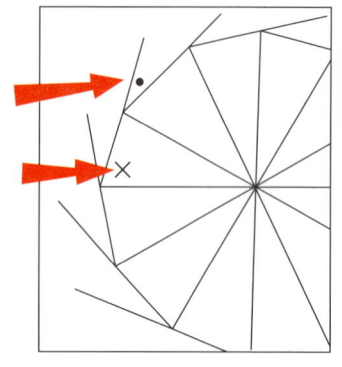

C9 (a) If you use the method of C8 but mark off every 10° round your angle measurer, how many sides will your polygon have?

  (b) What will each exterior angle be?

C10 (a) If you use the method of C8 to draw a regular polygon with 15 sides, how many degrees must you keep marking off round the angle measurer?

  (b) What will each exterior angle be?

  (c) What will each interior angle be?

C11 These are instructions to draw a certain kind of regular polygon on a computer.

(a) What kind of regular polygon is it?

(b) Write a version of the instructions that will draw a pentagon.

(c) Write a version that will draw an enlargement with scale factor 2 of the pentagon in (b).

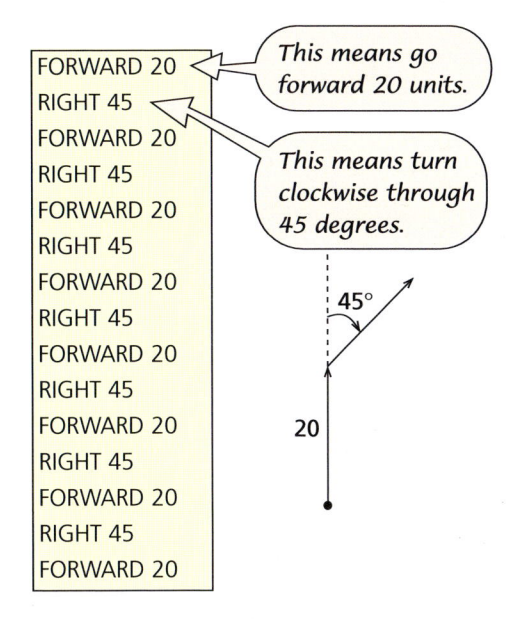

```
FORWARD 20
RIGHT 45
FORWARD 20
RIGHT 45
FORWARD 20
RIGHT 45
FORWARD 20
RIGHT 45
FORWARD 20
RIGHT 45
FORWARD 20
RIGHT 45
FORWARD 20
RIGHT 45
FORWARD 20
```

This means go forward 20 units.

This means turn clockwise through 45 degrees.

C12 The course for a sailing race is clockwise round a regular hexagon PQRSTU.

The first 'leg' of the race, P to Q, is on a bearing of 060°.

(a) What angle does a boat turn through when it gets to Q?

What are the bearings of these?

(b) The second leg of the race, from Q to R

(c) The third leg, R to S

(d) The fourth leg, S to T

(e) The fifth leg, T to U

(f) The sixth leg, U to P

C13 The course for a sailing race is clockwise round a square ABCD.

The first leg of the race, A to B, is on a bearing of 070°.

(a) What angle does a boat turn through when it gets to B?

What are the bearings of these?

(b) The second leg of the race, B to C

(c) The third leg, C to D

(d) The fourth leg, D to A

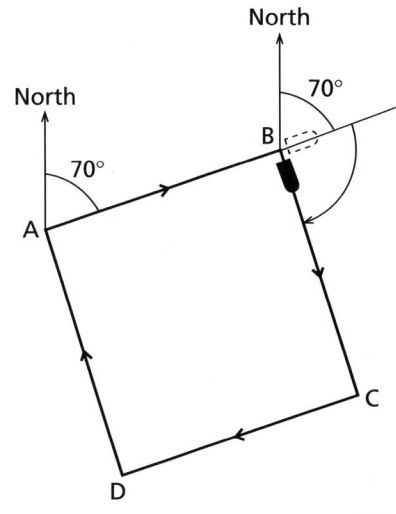

What progress have you made?

Statement

I can work out, and use, the sum of the interior angles of a polygon.

Evidence

1 What is the sum of the interior angles of a polygon with 11 sides?

2 What are the missing angles in these?
   (a) [polygon with angles 111°, 94°, 123°, ?, 140°, 132°]
   (b) [quadrilateral with angles 90°, 154°, 166°, ?]

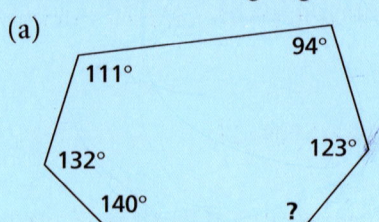

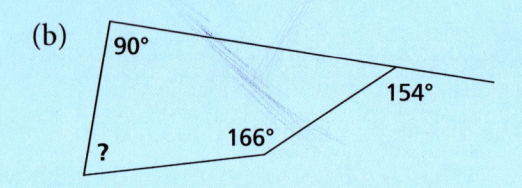

I know, and can use, the sum of the exterior angles of a polygon.

3 Calculate the missing angle.
[polygon with exterior angles 74°, 81°, 91°, 52°, ?]

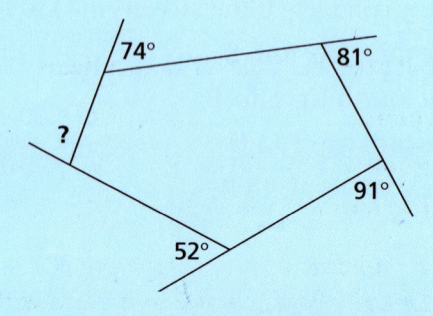

I can work with the angles of a regular polygon.

4 A regular polygon has 30 sides.
   (a) What is each exterior angle?
   (b) What is each interior angle?

5 Each interior angle of a regular polygon is 174°.
   (a) What is each exterior angle?
   (b) How many sides does the polygon have?

15 Using and misusing statistics

This is about ways of presenting statistical information.

The work will help you
- revise some ways of presenting information
- choose those that are most effective for a particular piece of work
- criticise the way in which statistical information has been gathered or presented

A Presenting data

A **time series graph** shows what happens to some value as time passes. Time usually goes on the across axis.

You can join points with straight lines, or with a smooth curve where it makes sense to do so.

A time series graph is good for noticing general trends as well as sudden changes.

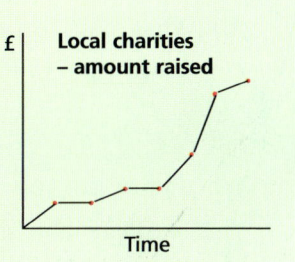

A1 This table gives the heights of two twins, Amy and Dan, at different ages.

Age in years	0	2	4	6	8	10	12	14	16	18
Amy's height (cm)	50	87	100	110	120	133	150	167	172	173
Dan's height (cm)	60	92	110	123	136	148	152	161	178	180

(a) Draw suitable axes on graph paper.
Plot points for Amy and join them with a smooth curve. Label it 'Amy'.
On the same axes, draw and label a curve for Dan the same way.
(b) Who was taller at age 11?
(c) When were Amy and Dan the same height?
(d) During what period was Amy taller than Dan?
(e) Describe the general features of Amy's growth from birth to age 18.
(f) Describe Dan's growth, commenting on how it differs from Amy's.

- In which of the problems on page 102 might a time series graph be useful?
- How would you use it?

A **two-way table** can be a good way to show how many things have – or do not have – certain properties (rather than measurements of some sort).

A2 Some pupils were asked what they saw when they looked at this picture. Some saw a candlestick (C). Others saw two faces (F).

Boy(B) or girl(G)	B	B	G	B	G	G	G	B	B	G	B	G
Picture	C	C	F	C	C	F	F	C	C	C	C	F

This records a girl who saw faces.

(a) Use the data to complete this two-way table.

(b) What main features does it show about this group of pupils?

	Girls	Boys	Total
Candlestick			
Faces			
Total			

A3 Whether or not you can smell certain flowers is thought to depend on your genes.

This two-way table shows the results of a survey of some adults with one such flower.

(a) Copy the table and add row and column totals.

(b) What percentage of the males could smell the flower? How does this compare with the females?

	Can smell	Cannot smell
Male	59	29
Female	47	15

- In which of the problems on page 102 might a two-way table be useful?
- How would you use it?

Summary statistics are values like the mean, median and range.
The mean and median give you a general idea about the size of values in a set of data.
The range tells you how spread out or variable the values are.

A4 Find the mean, median and range of this set of lengths.

13.6 m   12.2 m   14.1 m   12.9 m   16.1 m   13.2 m   16.8 m   13.5 m   13.0 m   17.6 m   13.6 m

- In which of the problems on page 102 might summary statistics be useful?
- How would you use them?

A grouped frequency bar chart can be a good way of summarising a lot of values. Surprising facts may be revealed that you could not 'take in' by looking at the original data.

It is often best to make a tally table first. Then draw the chart using the frequencies you got from the tally table.

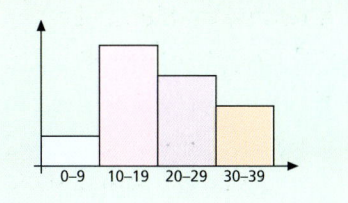

A5 Each of these data sets contains the estimated values (in thousands of pounds) of a random sample of 48 homes in a particular council ward (voting area).

Ward A

119	76	223	138	119	118	105	95	95	100	233	199
145	150	210	135	82	77	145	82	71	150	69	134
100	92	250	225	134	95	215	95	145	119	145	105
92	229	86	145	134	231	238	89	68	138	88	119

Ward B

135	65	185	150	140	129	129	111	111	129	199	180
150	165	180	149	72	72	165	83	62	168	35	148
129	95	201	185	145	115	185	105	165	130	165	129
103	185	85	150	142	192	205	95	27	150	87	136

(a) Choosing suitable intervals, draw a grouped frequency chart for each ward.
(b) Describe the general differences in the values of homes in the two wards.
(c) Find the mean and range of the values in the two wards.
    How useful are they for making a comparison in this case?

- Which of the problems on page 102 might a grouped frequency chart be useful for?
- How would you use it?

A **stem-and-leaf table** has the same effect as a grouped frequency bar chart. But it lets you still see all the original data, which is an advantage if you want to do further work on it.

```
3 | 2 7 7
2 | 4 5
1 | 3 8 9 9
0 | 2 2 3
```

Problems

You could investigate some of these by collecting your own data.
For others you would need to look in libraries or on the internet.

A
Sue writes with her right hand.
When she clasps her hands in front of her, she does it with her left thumb on top.
When she folds her arms, she does it with her right arm on top.
When she kicks a ball, she finds it most natural to use her right foot.
When she does a long jump, she pushes off from the ground with her right foot.

These are all examples of **limb dominance**.

Do different kinds of limb dominance tend to be related?
– for example, do most right-handers kick a ball with their right foot?

B
What has happened to people's consumption of beef in the last few years?
What has happened to their consumption of poultry?

C
How old are the pupils in your year group?

D
Do vegetarians avoid buying leather products?

E
How is the growth of a plant affected by keeping it in the shade?
Are different sorts of plant affected the same way?

F
What is the price of items that pupils spend their own money on?

G
Do the heights of girls vary more than the heights of boys?
Is the result different for different age groups?

H
What are the main causes of death of people over seventy?
How do these compare with other age groups?
How have things changed compared with the past?

I
How well paid are different sorts of part-time jobs?

- What ways of displaying or summarising data (besides those on pages 99 to 101) would be effective for these problems?

B Misleading charts and pictures

- What could be misleading about each of these charts?

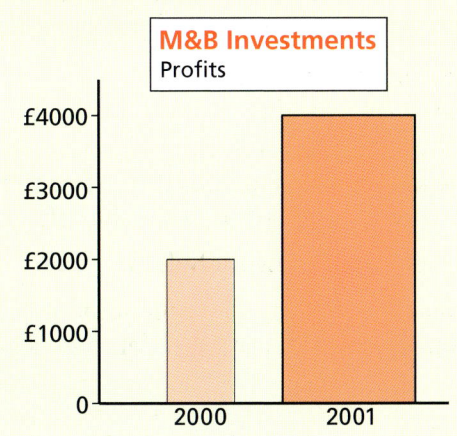

YUK COMPUTERS Sales figures
(bar chart showing Last month: 2500, This month: 3000; y-axis from 2000 to 3000)

M&B Investments Profits
(bar chart showing 2000: £2000, 2001: £4000)

- Roughly what percentage of the pie do you think is meat, gravy or crust?

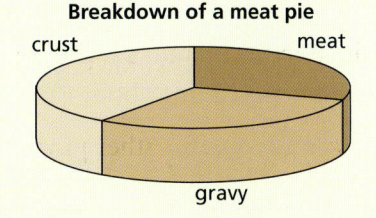

Breakdown of a meat pie (3D pie chart with sections labelled crust, meat, gravy)

- This advertisement about the National Lottery appeared in newspapers.

   Why might it be described as misleading?

   How would you improve it?

THE NATIONAL LOTTERY (United Kingdom) £2.3 billion
DAI-ICHI KANGYO BANK LOTTERY (Japan) £2.0 billion
ONLAE (Spain) £1.3 billion
LA FRANÇAISE DES JEUX (France) £1.0 billion

LOTTERIES CONTRIBUTING MOST TO GOOD CAUSES AND GOVERNMENT DUTY TAXES

Source: La Fleur's Lottery World, The Worldwide Lottery Efficiency Study 1996, based on the top 30 lotteries worldwide ranked by government profit

These questions can be tackled by pairs working together.

B1 The next two charts could mislead in some way.

Find what could be misleading about each chart and suggest how you might make it less misleading.

(a) **Workers with at least a minimum recognised qualification**

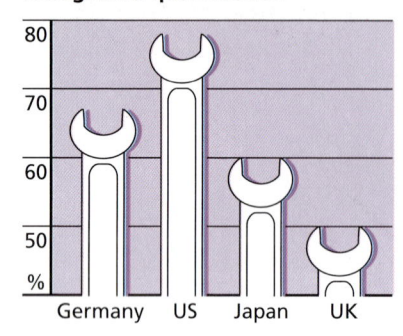

(b) **Percentage of 3–5 year olds receiving some childcare in Europe**

Country	Percentage
France	95%
Belgium	95%
Italy	88%
Denmark	87%
Spain	66%
Greece	62%
West Germany	60%
Ireland	52%
Netherlands	50%
Luxembourg	48%
United Kingdom	44%
Portugal	25%

B2 Find what could be misleading about each of these.

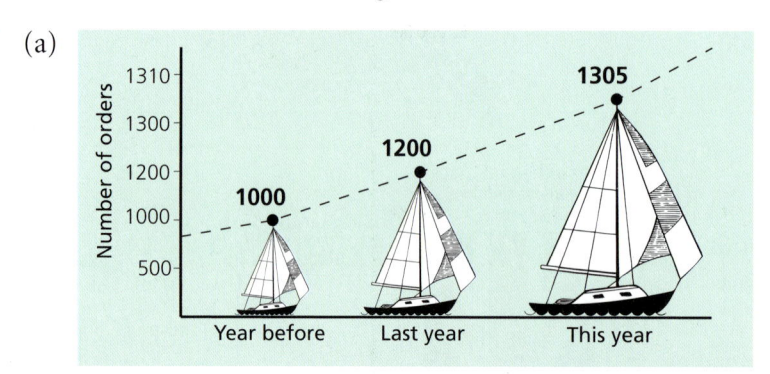

(a) [Chart showing number of orders with sailboats: Year before 1000, Last year 1200, This year 1305]

(b) [Hot air balloons: Last year we flew 36 times; This year we flew 54 times]

'Up, up and away' balloon flights report a 50% increase in flights

B3 These 3-D pie charts all show the same data about how Dwayne spends his day.
Do they all give the same impression?
How easy is it to estimate the percentage of time Dwayne spends
on each activity (or inactivity!)?

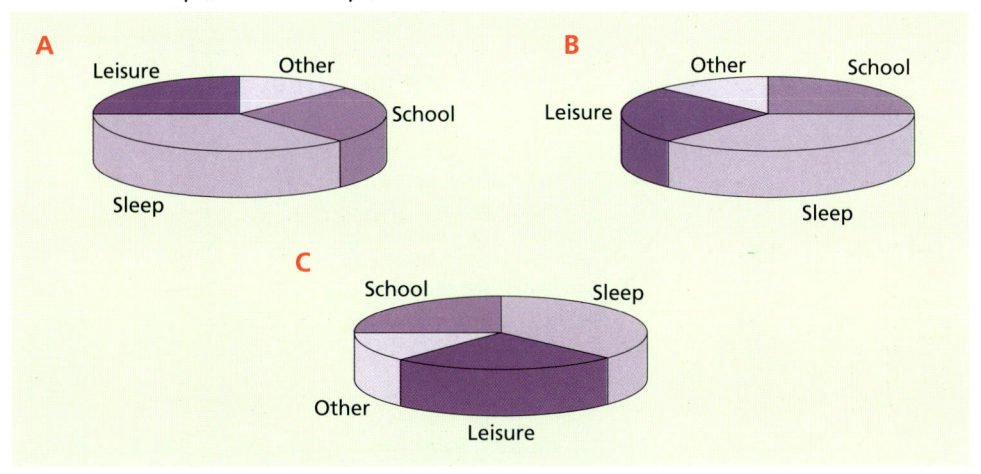

B4 Look at the two graphs here.
Which of the two businesses is growing faster?

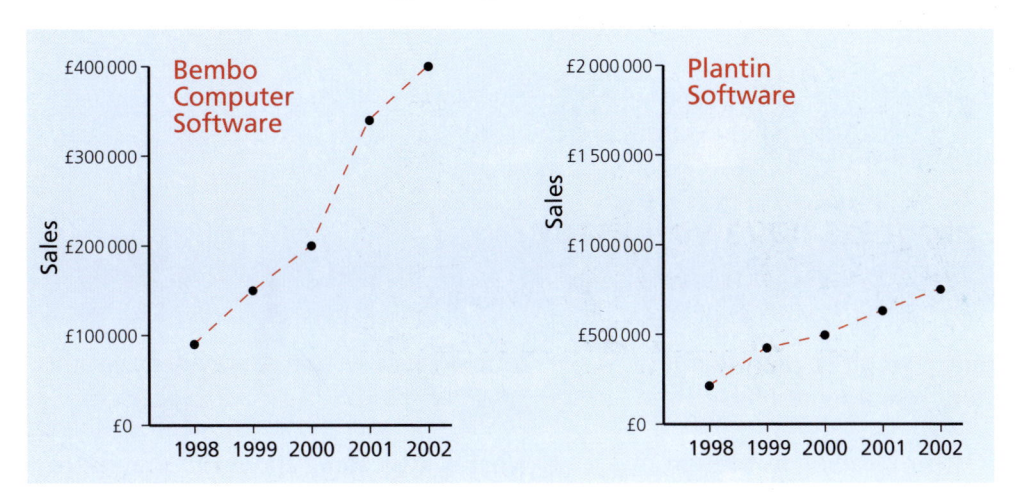

B5 These charts come from a phone company's report to customers.
Find what could be misleading about each one.

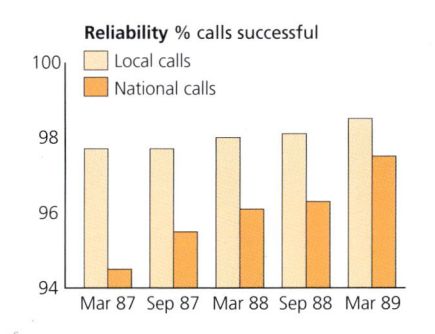

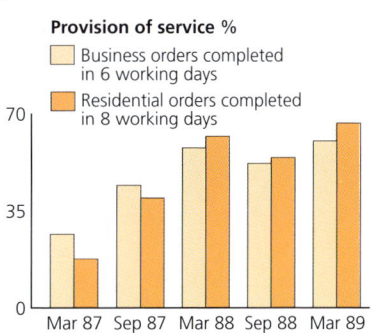

B6 Look at this headline and graph. Is the headline fair?

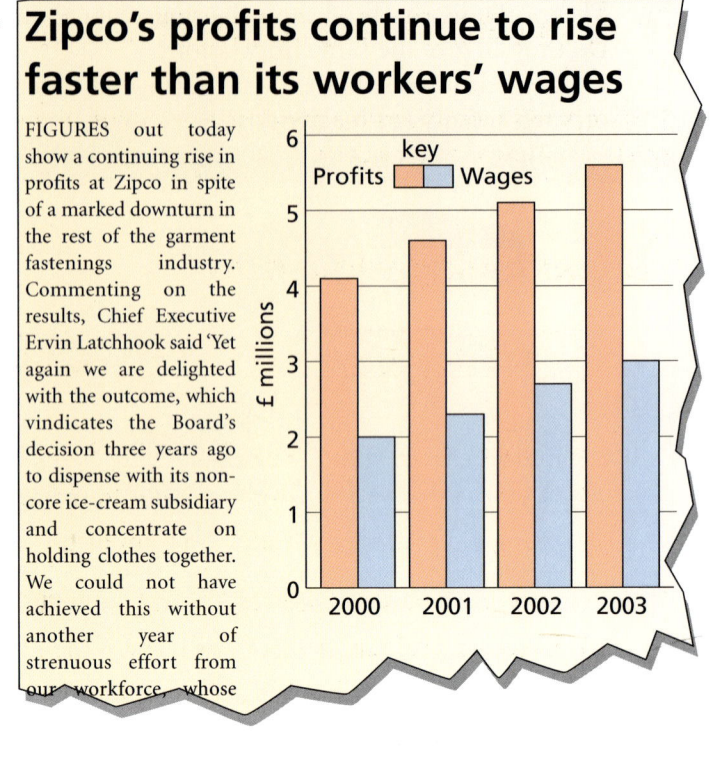

Zipco's profits continue to rise faster than its workers' wages

FIGURES out today show a continuing rise in profits at Zipco in spite of a marked downturn in the rest of the garment fastenings industry. Commenting on the results, Chief Executive Ervin Latchhook said 'Yet again we are delighted with the outcome, which vindicates the Board's decision three years ago to dispense with its non-core ice-cream subsidiary and concentrate on holding clothes together. We could not have achieved this without another year of strenuous effort from our workforce, whose

What progress have you made?

Statement

I can choose a graph or chart that is effective for presenting results.

I can identify misleading features in graphs, charts and diagrams.

Evidence

Your work in section A is evidence of this.

1 What is misleading about this graph?

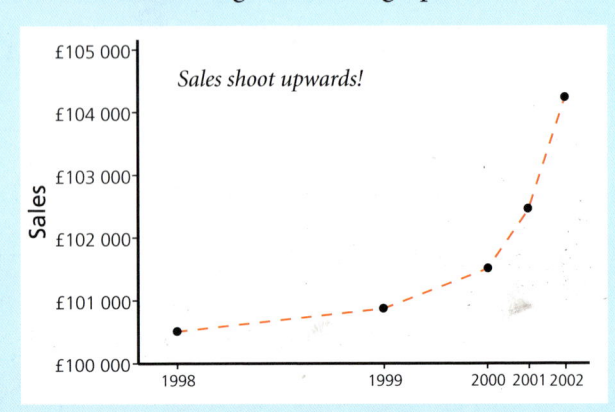

Sales shoot upwards!

16 Linear sequences

This work will help you find a rule for the *n*th term of any sequence that goes up or down in equal steps.

A Continuing sequences

> 1, 2, 4, ... ?

> 2, 4, 6, ... ?

A1 A sequence of numbers begins 7, 11, 15, 19, 23, ...

(a) Describe a rule to go from one term to the next.

(b) Using your rule, what is the 8th term of this sequence?

A2 A sequence of numbers begins 5, 6, 8, 11, 15, 20, ...

(a) Describe a rule to go from one term to the next.

(b) Using your rule, what is the 7th term of this sequence?

A3 For each of the following sequences,

- describe a rule to go from one term to the next
- find the 8th term

(a) 9, 14, 19, 24, 29, ...   (b) 28, 26, 24, 22, 20, ...

(c) 3, 6, 12, 24, 48, ...   (d) 1, 2.5, 4, 5.5, 7, ...

(e) 1, 3, 9, 27, 81, ...   (f) 5, 10, 15, 20, 25, ...

A4 Linear sequences go up or down in equal steps.

Which of these sequences are linear?

A 4, 10, 16, 22, 28, 34   **B** 3, 6, 9, 15, 24, 39   **C** 3, 4, 5, 7, 9, 12

D 4, 8, 16, 32, 64   **E** 17, 14, 11, 8, 5   **F** 3, 3.5, 4, 4.5, 5

A5 You need sheet 264.

Linear sequences can be found in the grid of numbers.
Two are shown on the diagram.

Find as many linear sequences as you can that have four terms or more.

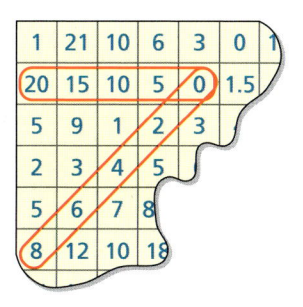

1	21	10	6	3	0	1
20	15	10	5	0	1.5	
5	9	1	2	3		
2	3	4	5			
5	6	7	8			
8	12	10	19			

A6 Copy and complete each of these linear sequences.

(a) 2, 9, 16, ___ , 30, ___

(b) 4, 6, ___, ___, 12, ___

(c) 10, ___, ___, 25, 30

(d) ___, 8, ___, ___, 17, ___

*A7 Solve these sequence puzzles.
All the sequences are linear.

Puzzle 1	Puzzle 2	Puzzle 3	Puzzle 4
My differences are 6. My 2nd term is 9. My 6th term is bigger than my 5th term. Find my 5th term.	My 1st term is 0. My 8th term is 21. Find my 9th term.	My 3rd term is 5. My 9th term is 29. Find my 6th term.	My 4th term is 36. My 20th term is 4. Find my 8th term.

B Sequences from rules

The rule for the nth term of a particular sequence is $2n + 9$.
We can use a table to show the first few terms of the sequence.

Term numbers (n)	1	2	3	4	5	6	...
Terms of the sequence ($2n + 9$)	11	13	15	17	19	21	...

The difference is 2 each time so this is a linear sequence.

B1 The nth term of a sequence is $n + 5$.

(a) Copy and complete the table below to show the first six terms of the sequence.

Term numbers (n)	1	2	3	4	5	6	...
Terms of the sequence ($n + 5$)	6						

(b) What are the differences for this sequence?
Is it a linear sequence?

B2 The nth term of a sequence is $3n - 2$.

(a) Copy and complete the table below to show the first six terms of the sequence.

Term numbers (n)	1	2	3	4	5	6	...
Terms of the sequence ($3n - 2$)							

(b) What are the differences for this sequence?
Is it a linear sequence?

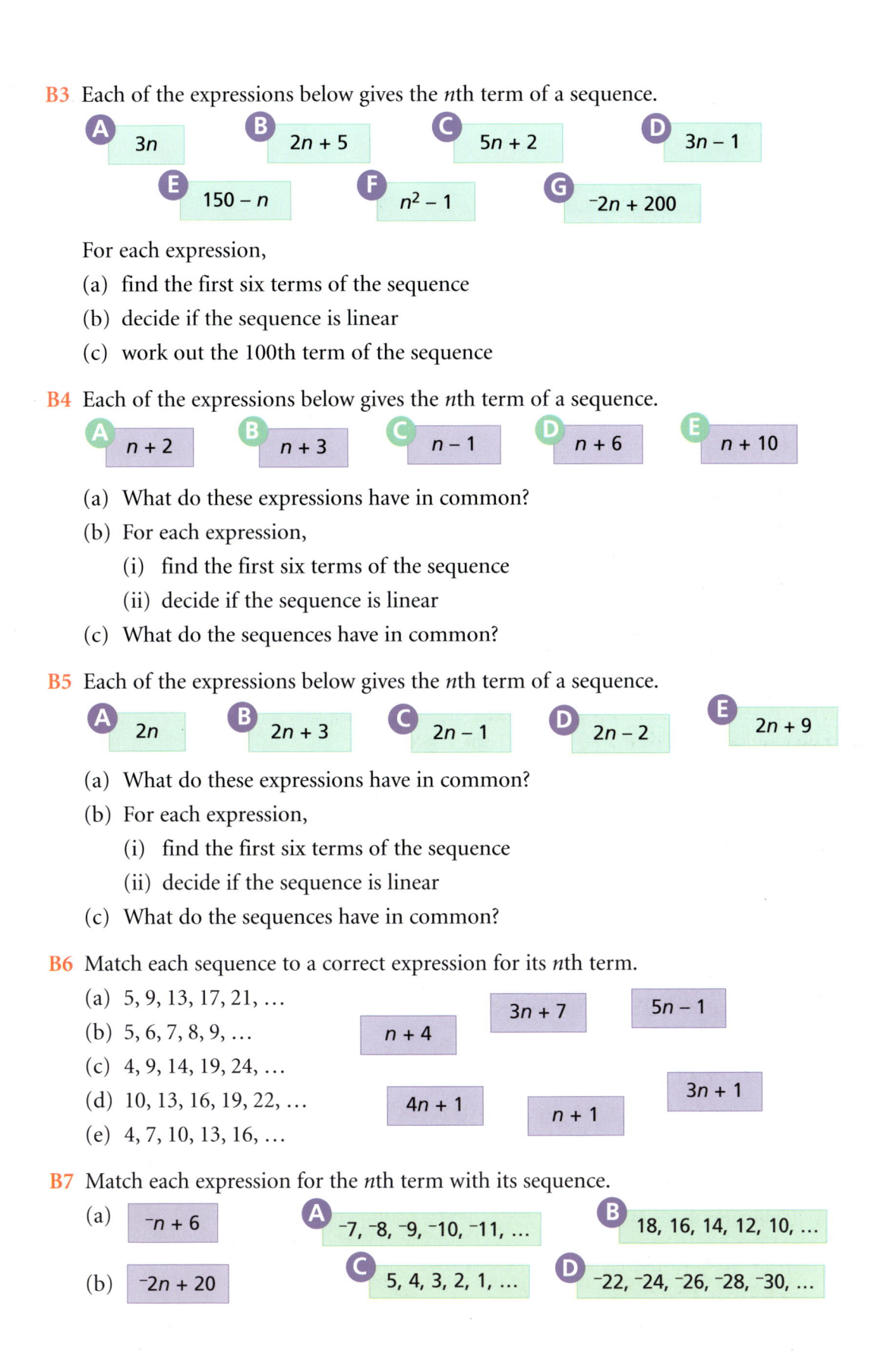

B3 Each of the expressions below gives the *n*th term of a sequence.

- **A** $3n$
- **B** $2n + 5$
- **C** $5n + 2$
- **D** $3n - 1$
- **E** $150 - n$
- **F** $n^2 - 1$
- **G** $-2n + 200$

For each expression,
(a) find the first six terms of the sequence
(b) decide if the sequence is linear
(c) work out the 100th term of the sequence

B4 Each of the expressions below gives the *n*th term of a sequence.

- **A** $n + 2$
- **B** $n + 3$
- **C** $n - 1$
- **D** $n + 6$
- **E** $n + 10$

(a) What do these expressions have in common?
(b) For each expression,
   (i) find the first six terms of the sequence
   (ii) decide if the sequence is linear
(c) What do the sequences have in common?

B5 Each of the expressions below gives the *n*th term of a sequence.

- **A** $2n$
- **B** $2n + 3$
- **C** $2n - 1$
- **D** $2n - 2$
- **E** $2n + 9$

(a) What do these expressions have in common?
(b) For each expression,
   (i) find the first six terms of the sequence
   (ii) decide if the sequence is linear
(c) What do the sequences have in common?

B6 Match each sequence to a correct expression for its *n*th term.
(a) 5, 9, 13, 17, 21, …
(b) 5, 6, 7, 8, 9, …
(c) 4, 9, 14, 19, 24, …
(d) 10, 13, 16, 19, 22, …
(e) 4, 7, 10, 13, 16, …

$3n + 7$ $5n - 1$ $n + 4$ $4n + 1$ $n + 1$ $3n + 1$

B7 Match each expression for the *n*th term with its sequence.
(a) $-n + 6$
(b) $-2n + 20$

- **A** −7, −8, −9, −10, −11, …
- **B** 18, 16, 14, 12, 10, …
- **C** 5, 4, 3, 2, 1, …
- **D** −22, −24, −26, −28, −30, …

C Increasing linear sequences

One way to find the nth term for the linear sequence: 8, 11, 14, 17, 20, … is shown below.

Make a table with space for working.

Term numbers (n)	1	2	3	4	5	…
Terms of the sequence	8	11	14	17	20	…

The sequence 8, 11, 14, 17, 20, … goes up in 3s so the rule begins $3n$ …

Write the sequence for $3n$.

Term numbers (n)	1	2	3	4	5	…
$3n$	3	6	9	12	15	
Terms of the sequence	8	11	14	17	20	…

To get from 3 to 8 you can add 5.

Adding 5 works for all terms.

Term numbers (n)	1	2	3	4	5	…
$3n$	3 +5	6 +5	9 +5	12	15	
Terms of the sequence	8	11	14	17	20	…

So the rule for the nth term is $3n + 5$.

C1 For each of the following linear sequences,
- find an expression for the nth term
- calculate the 50th term in the sequence

  (a) 9, 12, 15, 18, 21, …   (b) 1, 6, 11, 16, 21, …
  (c) 7, 9, 11, 13, 15, 17, …   (d) 1, 7, 13, 19, 25, 31, …
  (e) 10, 11, 12, 13, 14, …   (f) 1, 5, 9, 13, 17, …

C2 A pupil has tried to find the nth term of a linear sequence.

8, 10, 12, 14, 16, …
nth term is $n + 2$ ✗

  (a) Explain the mistake you think he has made.
  (b) Find a correct expression for the nth term of this sequence.

C3 This sequence of dolphins continues to the right.

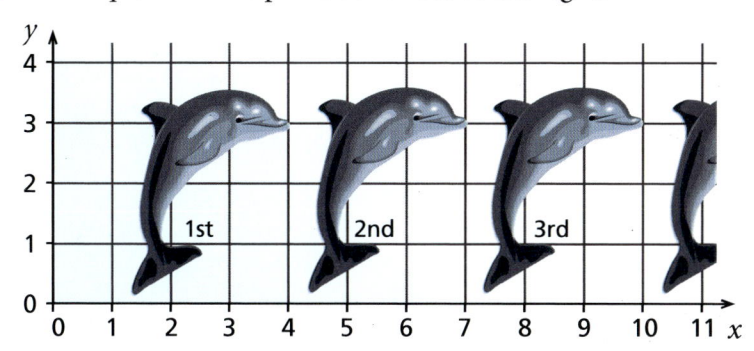

(a) What are the coordinates of the 1st dolphin's nose?

(b) Give the coordinates for the 2nd dolphin's nose.

(c) Work out the coordinates for

(i) the 4th dolphin's nose   (ii) the 10th dolphin's nose

(d) Show coordinates for the first ten dolphins in a table like this.

Dolphin coordinates table

Dolphin (n)	x	y
1		
2		
3	10	3
4		
5		

(e) Give the coordinates of the nth dolphin's nose.

(f) What are the coordinates of the 30th dolphin's nose?

*C4 This sequence of birds continues up to the right.

(a) List the coordinates of the tips of the birds' tails.

(b) Put your coordinates in a table.

(c) Give the coordinates of the nth bird's tail tip.

(d) What are the coordinates of the 100th bird's tail tip?

(e) Try to find the coordinates of the nth bird's wing tip.

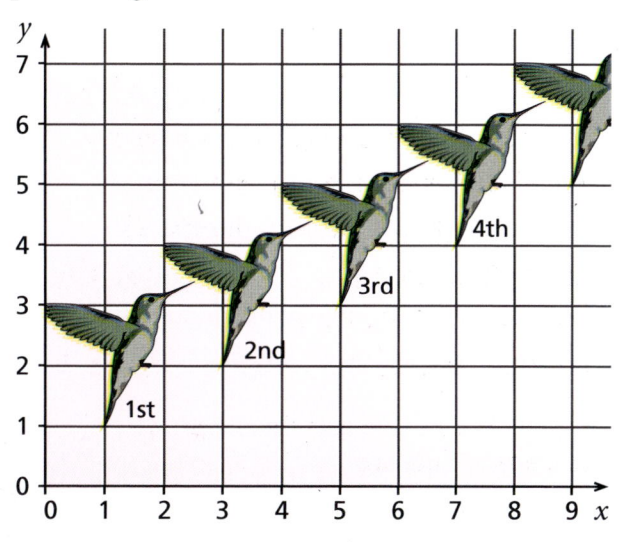

D Decreasing linear sequences

One way to find the *n*th term for the linear sequence: 50, 48, 46, 44, 42, … is shown below.

Make a table with space for working.

Term numbers (*n*)	1	2	3	4	5	…
Terms of the sequence	50	48	46	44	42	…

The sequence 50, 48, 46, 44, 42, … goes down in 2s so the rule begins ⁻2n …

Write the sequence for ⁻2n.

Term numbers (*n*)	1	2	3	4	5	…
⁻2*n*	⁻2	⁻4	⁻6	⁻8	⁻10	
Terms of the sequence	50	48	46	44	42	…

To get from ⁻2 to 50 you can add 52.

Adding 52 works for all terms.

Term numbers (*n*)	1	2	3	4	5	…
⁻2*n*	⁻2	⁻4	⁻6	⁻8	⁻10	
	+ 52	+ 52	+ 52	+ 52	+ 52	
Terms of the sequence	50	48	46	44	42	…

So the rule for the nth term is ⁻2n + 52. This is usually written 52 − 2n.

D1 For each of the following linear sequences,
- find an expression for the *n*th term
- calculate the 20th term in the sequence

(a) 98, 96, 94, 92, 90, …
(b) 70, 67, 64, 61, 58, …
(c) 49, 48, 47, 46, 45, …
(d) 95, 90, 85, 80, 75, …
(e) 86, 82, 78, 74, 70, …
(f) 47, 44, 41, 38, 35, …

D2 This sequence of sharks continues downwards.

(a) What are the coordinates of the 1st shark's nose?

(b) Give the coordinates for the 3rd shark's nose.

(c) Work out the coordinates for
  (i) the 4th shark's nose
  (ii) the 6th shark's nose

(d) Show coordinates for the first ten sharks in a table like this.

Shark nose coordinates table		
Shark (n)	x	y
1		
2	1	9
3		
4		
5		

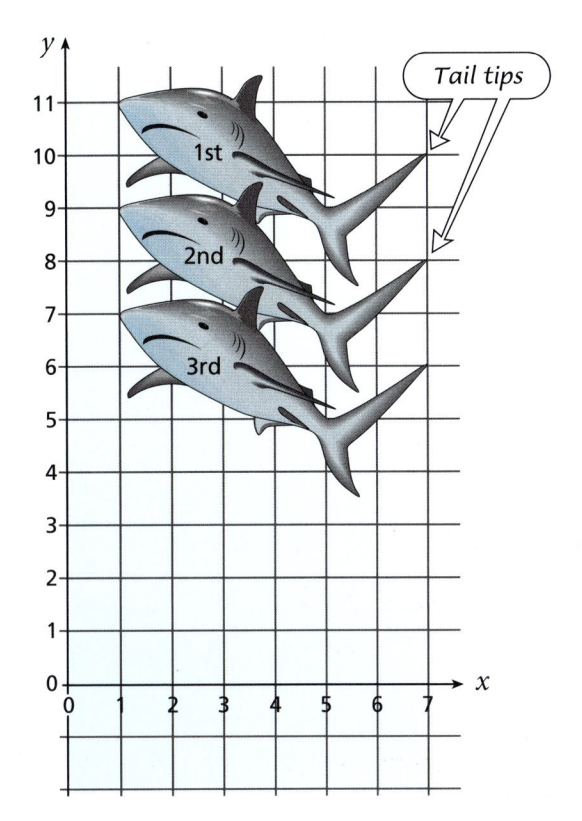

(e) Give the coordinates of the nth shark's nose.

(f) Make a table for the coordinates of the shark's tail tips.

(g) What are the coordinates of the nth shark's tail tip?

(h) Give the coordinates of the 20th shark's tail tip.

***D3** This sequence of penguins continues down to the right.

(a) Make a table for the coordinates of the penguins' beaks.

(b) Give the coordinates of the nth penguin's beak.

(c) What are the coordinates of the 50th penguin's beak?

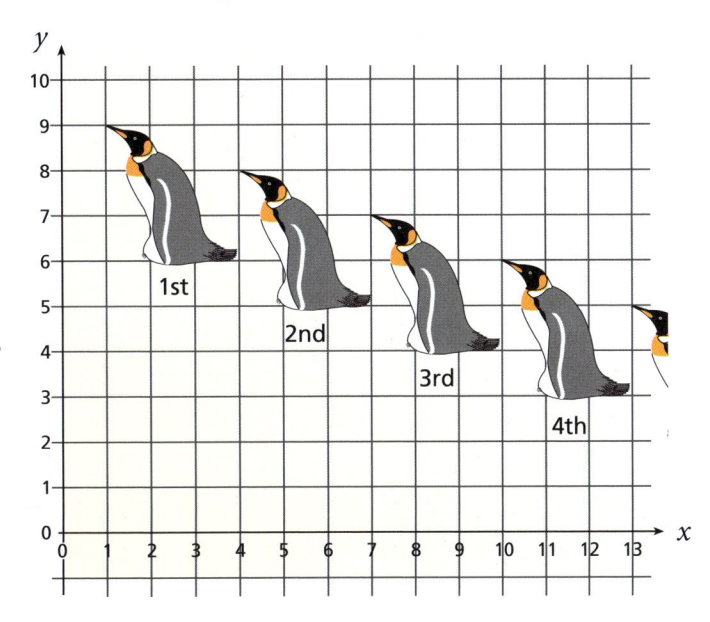

What progress have you made?

Statement

I can find and use rules to continue a variety of sequences.

Evidence

Here are some sequences.

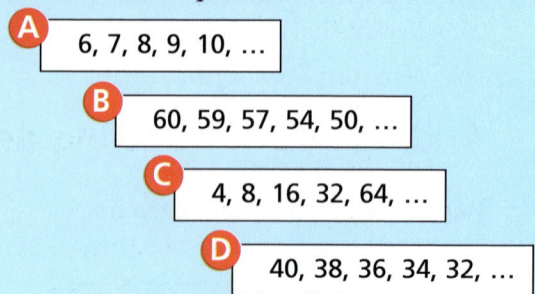

A 6, 7, 8, 9, 10, …

B 60, 59, 57, 54, 50, …

C 4, 8, 16, 32, 64, …

D 40, 38, 36, 34, 32, …

1 For each sequence above,
   (a) describe a rule to go from one term to the next
   (b) find the 10th term

I know when a sequence is an increasing or decreasing linear sequence.

2 Which of the above sequences is
   (a) an increasing linear sequence
   (b) a decreasing linear sequence

I can use a rule for the nth term of a linear sequence.

3 The nth term of a sequence is $5n - 2$.
   (a) Find the first six terms of this sequence.
   (b) What is the 20th term?

4 The nth term of a sequence is $28 - n$.
   (a) Find the first four terms of this sequence.
   (b) What is the 10th term?

I can find and use a rule for the nth term of a linear sequence.

5 For each of the following linear sequences,
   - find an expression for the nth term
   - calculate the 25th term
   (a) 11, 17, 23, 29, 35, …
   (b) 2, 6, 10, 14, 18, 22, …

6 For each of the following linear sequences,
   - find an expression for the nth term
   - calculate the 30th term
   (a) 39, 38, 37, 36, 35, …
   (b) 150, 146, 142, 138, 134, …

17 Decimals

This work will help you
- use further mental and written methods for calculating with decimals
- solve problems involving decimals

A Adding and subtracting decimals – revision

A1 Work these out in your head.
(a) 8.2 + 10.9 (b) 11.0 – 0.8 (c) 0.07 + 5.25 (d) 10.06 + 2.26
(e) 3.04 + 0.68 (f) 70.45 – 2.21 (g) 52.6 – 10.7 (h) 30 – 5.74

A2 In your head, work out the change from £20.00 for each of these.
(a) £1.50 (b) 78p (c) £12.95 (d) £14.27 (e) £7.08

A3 Work out in your head
(a) the total weight of these cats
(b) the difference in their weights

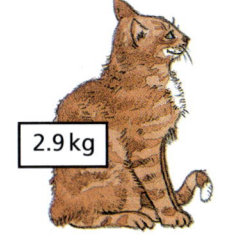

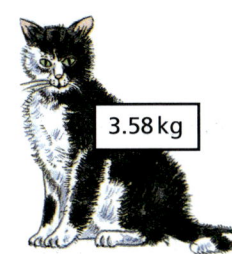

2.9 kg   3.58 kg

A4 These are the heights in metres of a group of six-year-old children.
0.96 1.08 1.12 0.87 1.14 1.02 0.91
In your head, work out the range of their heights.

Written addition and subtraction of decimals

Remember to line up the decimal points. You may need to put in zeros.

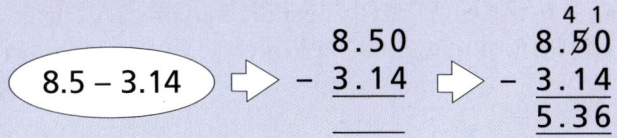

$$6.8 + 1.32 \Rightarrow \begin{array}{r} 6.80 \\ +1.32 \\ \hline 8.12 \\ {}_1 \end{array}$$

$$8.5 - 3.14 \Rightarrow \begin{array}{r} 8.50 \\ -3.14 \\ \hline \end{array} \Rightarrow \begin{array}{r} {}^{4}8.\overset{1}{\cancel{5}}0 \\ -3.14 \\ \hline 5.36 \end{array}$$

A5 Copy these, filling in the missing digits.

(a)   1 . 6
    + ■ . 5 2
    ─────────
      3 . ■ ■

(b)   ■ . 8
    + ■ 6 . ■ 6
    ─────────
      3 0 . 0 ■

(c)   5 . 4 3
    – 2 . ■ ■
    ─────────
      ■ . 8 2

(d)   3 . ■ 1
    – 1 . 1 ■
    ─────────
      ■ . 4 8

A6 Use written methods to calculate these.
   (a) 129.6 + 3.85   (b) 268.58 − 10.2   (c) 57.9 + 1.803   (d) 347 − 59.3
   (e) 30.4 − 6.202   (f) 1.06 + 19.535 + 206.8   (g) 302.4 + 18.203 − 1.42

A7 Nicola tried to work out 138.27 + 64.5 on her calculator and got the answer 82.77. Which key did she not press properly?

Adding and subtracting quantities in different metric units

Before calculating, you must convert so all the quantities are in the same units.

175 g of ice cream is used from a tub that contained 2.5 kg. How much is left?

$175\,g = 0.175\,kg$

```
  2.500
− 0.175
  2.325
```

So 2.325 kg of ice cream is left.

(Or you could convert 2.5 kg to grams and work in grams.)

Linda is 1.69 m tall. Her son is 13 cm taller than her. How tall is her son?

$13\,cm = 0.13\,m$

```
  1.69
+ 0.13
  1.82
```

So Linda's son is 1.82 m tall.

(Or you could convert 1.69 m to centimetres and work in centimetres.)

A8 On 7 August 1995, Jonathan Edwards set a new world record of 18.16 m for the triple jump.
Ten minutes later, he broke his own record by jumping 13 cm further.
How far, in metres, did he jump the second time?

A9 A concrete post 2.05 m high is placed in a hole in the ground 38 cm deep.
What height of post shows above the ground?

A10 A 'two pint' carton of milk contains 1.14 litres when full.
How much milk is left after 125 ml of milk has been used?

A11 A certain model of car is 4.344 m long.
The manufacturer decides next year's version will be 2.8 cm longer.
How long, in metres, will next year's model be?

A12 Packets coming off a production line are labelled as containing 1 kg of sugar.
A quality controller takes a random sample of packets and weighs the sugar.
She finds the smallest weight is 0.962 kg and the largest is 1.067 kg.
   (a) How much less than 1 kg is is the packet with the smallest weight?
   (b) What is the range of the weights in the sample, in grams?

B Multiplying

B1 Work these out in your head.
(a) 1.5 × 4
(b) 0.9 × 7
(c) 2.7 × 2
(d) 1.6 × 4
(e) 0.02 × 3
(f) 0.04 × 6
(g) 0.12 × 5
(h) 1.1 × 9
(i) 2.08 × 7
(j) 3.15 × 4

B2 Choose numbers from the loop to make these multiplications correct.
You can use a number more than once.
(a) ■ × ■ = 2.4
(b) ■ × ■ = 0.15
(c) ■ × ■ = 0.48
(d) ■ × ■ = 2
(e) ■ × ■ = 0.3
(f) ■ × ■ = 3

Loop: 0.03, 0.6, 0.25, 8, 4, 5, 0.06

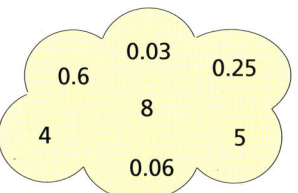

Multiplying a decimal by an integer: written methods

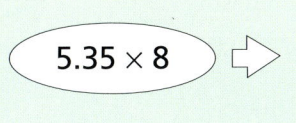

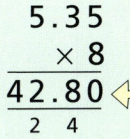

6.4 × 3 ⇒
```
  6.4
×   3
 19.2
  1
```

5.35 × 8 ⇒
```
   5.35
×     8
  42.80
   2 4
```
You could write this answer as 42.8.

30.26 × 18 ⇒
```
    30.26
×      18
   302.60
   242.08
   544.68
```

Check by rounding the given values.
30 × 20 = 600, so the answer is sensible.

B3 Work these out using a written method.
(a) 9.5 × 9
(b) 4.56 × 7
(c) 10.28 × 6
(d) 40.2 × 18
(e) 3.7 × 108
(f) 2.7 × 220
(g) 3.62 × 24
(h) 6.19 × 53

B4 Use a written method to calculate the total cost of each of these.
(a) 7 netball outfits at £45.50 each
(b) 11 pairs of soccer boots at £28.75 each
(c) 13 rugby outfits at £60.15 each
(d) 8 oars at £107.35 each
(e) 450 number bibs for marathon runners at £2.05 each

Multiplying a decimal by a decimal

You can think of 0.3×0.3 as $\frac{3}{10} \times \frac{3}{10}$, which equals $\frac{9}{100}$, or 0.09.
Similarly, $0.7 \times 0.9 = \frac{7}{10} \times \frac{9}{10} = \frac{63}{100} = 0.63$
$0.4 \times 0.06 = \frac{4}{10} \times \frac{6}{100} = \frac{24}{1000} = 0.024$

B5 For each multiplication,
- write it as a fraction multiplied by a fraction
- give the answer as a fraction
- convert the answer to a decimal

(a) 0.1×0.7 (b) 0.3×0.1 (c) 0.9×0.3 (d) 0.4×0.3 (e) 0.8×0.6
(f) 0.3×0.05 (g) 0.01×0.8 (h) 0.7×0.07 (i) 0.06×0.04 (j) $0.3 \times 0.5 \times 0.7$

B6 Write the decimals that are missing here.

(a) $0.4 \times ? = 0.36$ (b) $0.2 \times ? = 0.06$ (c) $0.5 \times ? = 0.04$

B7 Write decimal answers for these.

(a) $0.1 \times 0.2 \times 0.3$ (b) $0.2 \times 0.5 \times 0.6$ (c) $0.4 \times 0.5 \times 0.5$

Squares and cubes

$0.2^2 = 0.2 \times 0.2 = 0.04$
$0.5^3 = 0.5 \times 0.5 \times 0.5 = 0.125$

B8 Work these out without using a calculator.

(a) 0.6^2 (b) 0.5^2 (c) 0.1^2 (d) 0.8^2 (e) 0.2^3
(f) 0.9^3 (g) 0.1^3 (h) 0.7^3 (i) 0.03^2 (j) 0.02^3

Square roots and cube roots

Since $0.2^2 = 0.04$, the square root of 0.04 is 0.2 (we write $\sqrt{0.04} = 0.2$).
Since $0.5^3 = 0.125$, the cube root of 0.125 is 0.5 (we write $\sqrt[3]{0.125} = 0.5$).

B9 Experiment without a calculator to find each of these.
When you have recorded your result, you can check with a calculator.

(a) $\sqrt{0.81}$ (b) $\sqrt{0.16}$ (c) $\sqrt{0.0049}$ (d) $\sqrt[3]{0.027}$ (e) $\sqrt[3]{0.216}$

B10 Without a calculator, decide whether each of these is true or false.

(a) $\sqrt{0.36} = 0.06$ (b) $\sqrt{0.49} = 0.7$ (c) $\sqrt[3]{0.512} = 0.8$ (d) $\sqrt[3]{0.64} = 0.4$

Multiplying and dividing a decimal by a power of 10 – a reminder

Examples $8.25 \times 1000 = 8250$ $47.1 \div 100 = 0.471$ $0.059 \times 10\,000 = 590$

B11 Answer these without using a calculator.
(a) 2.38×100 (b) 302×1000 (c) 0.822×10 (d) $82.6 \div 100$
(e) $0.407 \div 10$ (f) $64.7 \times 10\,000$ (g) $7.52 \div 1000$ (h) $588 \div 10\,000$
(i) $0.0216 \div 10$ (j) 0.0418×1000 (k) 9050×1000 (l) $0.309 \div 100$

B12 Convert these to kilograms. (a) 4820 g (b) 150 g (c) 10.3 g

B13 Convert these to centimetres. (a) 1.2 m (b) 154 mm (c) 0.078 m

B14 Convert these to metres. (a) 4.2 km (b) 0.052 km (c) 9.2 cm

Multiplying a decimal by a decimal: written methods

2.31×4.5 ⇒ Split the numbers into place value parts, multiply separately then add …

×	2	0.3	0.01
4	8	1.2	0.04
0.5	1	0.15	0.005

```
   8
   1.2
   0.04
   1
   0.15
   0.005
  ------
  10.395
```

or treat the decimals as whole numbers, multiply then adjust.

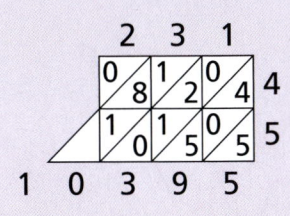

```
  2 3 1
 ┌─┬─┬─┐
 │0│1│0│
 │ 8 2 4│ 4
 │1│1│0│
 │ 0 5 5│ 5
 └─┴─┴─┘
1 0 3 9 5
```

```
     2 3 1      ← This is 100 times 2.31
   ×  4 5      ← and this is 10 times 4.5…
   -------
     9 2 4 0
     1 1 5 5
   -------
   1 0 3 9 5   ← so this is 100 × 10 times the true answer
                 (that's 1000 times the true answer).
```

So this answer must be divided by 1000 to get the true answer:

$10\,395 \div 1000 = 10.395$

Check by rounding the given values to the nearest whole number:

$2 \times 5 = 10$, so the answer is sensible.

- The number of decimal places in the answer equals the sum of the numbers of decimal places in the two given values. Why?

B15 Work these out without using a calculator. Check that each answer is sensible.
(a) 2.6×0.8 (b) 11.1×0.04 (c) 0.67×0.03 (d) 8.2×0.003 (e) 7.14×0.009
(f) 5.1×1.3 (g) 9.4×0.21 (h) 27×0.36 (i) 3.4×0.015 (j) 6.6×2.04

B16 Find the area of each of these rugs without using a calculator.
(a) 0.8 m, 1.4 m
(b) 2.3 m, 1.1 m

B17 Use the fact that $38 \times 27 = 1026$ to write the answer to each of these.
(a) 3.8×0.27 (b) 0.38×0.027 (c) 0.38×0.27 (d) 3.8×2.7 (e) 0.038×0.027

B18 Use the fact that $314 \times 23 = 7222$ to write the answer to each of these.
(a) 31.4×0.23 (b) 3.14×2.3 (c) 31.4×0.023 (d) 0.314×2.3 (e) 0.0314×0.23

B19 Work these out without using a calculator.
(a) 1.2^2 (b) 2.2^2 (c) 10.1^2 (d) 10.1^3 (e) 0.8^3

B20 Without a calculator, decide whether each of these is true or false.
(a) $\sqrt{1.21} = 1.1$ (b) $\sqrt[3]{1.331} = 0.11$ (c) $\sqrt{0.121} = 0.11$
(d) $\sqrt[3]{1.331} = 1.1$ (e) $\sqrt{0.0121} = 0.11$ (f) $\sqrt[3]{0.001331} = 0.11$

Not always bigger

When you look up 'multiply' in a dictionary the first definitions you see are usually like 'to increase the number of', 'to increase in number', 'to become more numerous'.

In mathematics too, multiplying a number often makes it bigger.
But not always.
- When does multiplying a number give a result that is smaller?

C Dividing

Dividing a decimal by an integer

Keep the decimal points one above the other.
Make sure your answer is sensible.

$14.7 \div 3 \Rightarrow 3\overline{)14.7} = 4.9$

C1 Work these out without a calculator.
(a) $1.2 \div 4$ (b) $3.96 \div 6$ (c) $10.08 \div 9$ (d) $48.3 \div 21$ (e) $0.195 \div 15$

Dividing by a decimal

You can change the decimal to a whole number by multiplying both given numbers by 10 or 100 or 1000 …

$$\frac{18}{0.3} \xrightarrow{\times 10} = \frac{180}{3} \qquad \frac{5.6}{0.24} \xrightarrow{\times 100} = \frac{560}{24} \qquad \frac{0.0012}{0.003} \xrightarrow{\times 1000} = \frac{1.2}{3}$$

Moving both decimal points the same number of places to the right has the same effect.

180. ÷ 03.   560. ÷ 024.   0001.2 ÷ 0003.

C2 Work these out without a calculator.
(a) $\frac{8}{0.4}$ (b) $\frac{14}{0.7}$ (c) $\frac{1.8}{0.09}$ (d) $\frac{120}{0.02}$ (e) $\frac{0.015}{0.005}$

C3 Work these out in your head.
(a) 66 ÷ 0.3 (b) 5.4 ÷ 0.9 (c) 24 ÷ 0.06 (d) 7.2 ÷ 0.008 (e) 560 ÷ 0.07

C4 Work these out using a written method.
(a) 19.2 ÷ 0.02 (b) 29.4 ÷ 0.4 (c) 0.3 ÷ 0.005 (d) 22.1 ÷ 1.3 (e) 3.1 ÷ 0.025

C5 Work these out using a written method. Keep going until you have two places of decimals in your answer then round to one decimal place.
(a) 38 ÷ 0.3 (b) 1.42 ÷ 0.07 (c) 15.7 ÷ 0.8 (d) 0.6 ÷ 0.011 (e) 0.04 ÷ 0.015

C6 Work these out giving your answer to two decimal places.
(a) 23 ÷ 0.6 (b) 7.4 ÷ 0.09 (c) 5.06 ÷ 0.7 (d) 0.1 ÷ 0.12 (e) 2.1 ÷ 1.6

C7 Work out each of these by a written method, checking that your answer is sensible.
(a) How many bottles of capacity 0.75 litre can be filled from a barrel holding 91.6 litres?
(b) How many lengths of cable 2.3 m long can be cut from a reel of cable 74.6 m long?

Not always smaller

We usually expect dividing a number to make it smaller. But this does not always happen.
- When does dividing a number give a result that is larger?

D Mixed questions

Do these questions without a calculator.

D1 Emily runs 6.8 km every day in July. How far is that altogether?

D2 Four of these calculations are wrong.
Find the wrong ones and do them correctly.

A 24.5 + 6.8

```
  24.5
+  6.8
  30.13
```

B 19 − 4.2

```
  19.0
−  4.2
  14.8
```

C 4.8 × 9

```
  4.8
×   9
 43.2
   7
```

D 13.32 + 2.8

```
  13.32
+  2.8
  13.60
```

E 20.3 ÷ 7

```
    2.9
  _____
7)20.⁶3
```

F 132 × 1.4

```
    132
  ×  1.4
   1320
    528
  184.8
```

G 13.05 ÷ 9

```
    1.05
  _____
9)13.⁴05
```

H 1.43 × 0.205

```
    1.43
× 0.205
  28600
    715
 29.315
```

D3 This shows a picture in a frame.
Work these out.

(a) The height of the picture
(b) The width of the picture
(c) The perimeter of the frame
(d) The perimeter of the picture
(e) The area of the picture
(f) The area of the frame

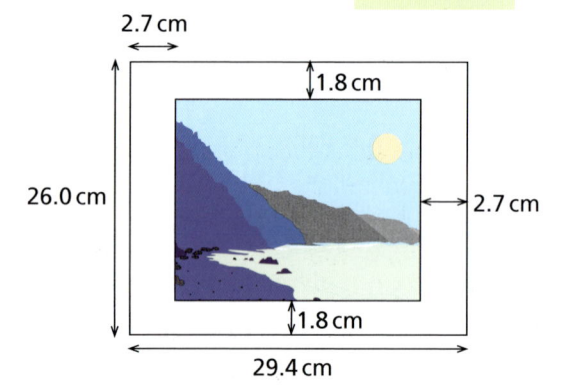

(Picture dimensions: 2.7 cm top margin, 1.8 cm, 26.0 cm, 2.7 cm, 1.8 cm, 29.4 cm)

D4 A section of a bridge is 32.848 m long when the temperature is 3°C.
It expands by 0.4 mm for every degree rise in temperature.
How long is the section when the temperature is 18°C?

D5 A recipe for coral pink paint says 'mix red and white in the ratio 4 : 11'.

(a) If you use 3.6 litres of red paint, how much white is needed?
(b) How much of each colour is needed to make 6 litres of coral pink?

D6 Use this conversion table to find the
metric equivalents of the quantities below.
Give your answers to a sensible degree of accuracy.

(a) $12\frac{1}{2}$ inches
(b) $5\frac{3}{4}$ pounds
(c) 86 gallons
(d) 375 feet
(e) $2\frac{1}{4}$ miles
(f) $6\frac{3}{4}$ pints

Imperial	Metric
1 pound	0.45 kg
1 mile	1.6 km
1 inch	2.5 cm
1 foot	0.3 m
1 pint	0.57 litres
1 gallon	4.5 litres

D7 Dave has 25 crates of machine parts to deliver to a factory.
Each crate weighs 0.7 tonne.
The maximum safe load for his lorry is 2.5 tonnes.
How many trips must he make to the factory to deliver all the crates safely?

D8 (a) Work out the perimeter of this shape.

(b) If the shape is enlarged by scale factor 3, what is the perimeter of the new shape?

D9 The seven sectors on this spinner are all the same size.

(a) Copy and complete this table to show the probability of getting each colour if the spinner is fair.

Colour	Probability as a fraction	Probability as a decimal (to 2 d.p.)
Blue		
Red		
Yellow		

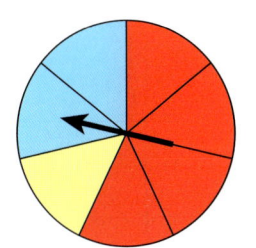

(b) Sophie spins the spinner 40 times.
She gets red 27 times and yellow 3 times.
Calculate the relative frequency of red and the relative frequency of yellow (these are estimates of their probability).
Give your answers as decimals to 2 d.p.

(c) Compare the relative frequencies from Sophie's experiment with your table.
Describe any differences.
Does her experiment prove that the spinner is not fair?

D10 Without doing the calculations, put these in order of size of answer, smallest first.

13×0.4 $13 \div 0.4$ 13×0.8 $13 \div 0.8$

D11 Find the missing numbers here.

(a) $48.5 \div ? = 0.0485$ (b) $4.8 \div ? = 0.6$ (c) $5.4 \div ? = 0.09$

D12 A piece of cheese weighing 0.23 kg costs £1.87.
What is the price per kilogram?

E Recurring decimals

You have met some recurring decimals that are equivalent to fractions, for example, $0.3333\ldots = \frac{1}{3}$, $0.1111\ldots = \frac{1}{9}$, $0.142\,857\,142\,857\,142\,857\ldots = \frac{1}{7}$

In fact **every** recurring decimal is equivalent to some exact fraction.
If you are given a recurring decimal, a chart like the one on the opposite page may help you find the fraction.

> **Example**
>
> Find the fraction that $0.727\,272\ldots$ is equivalent to.

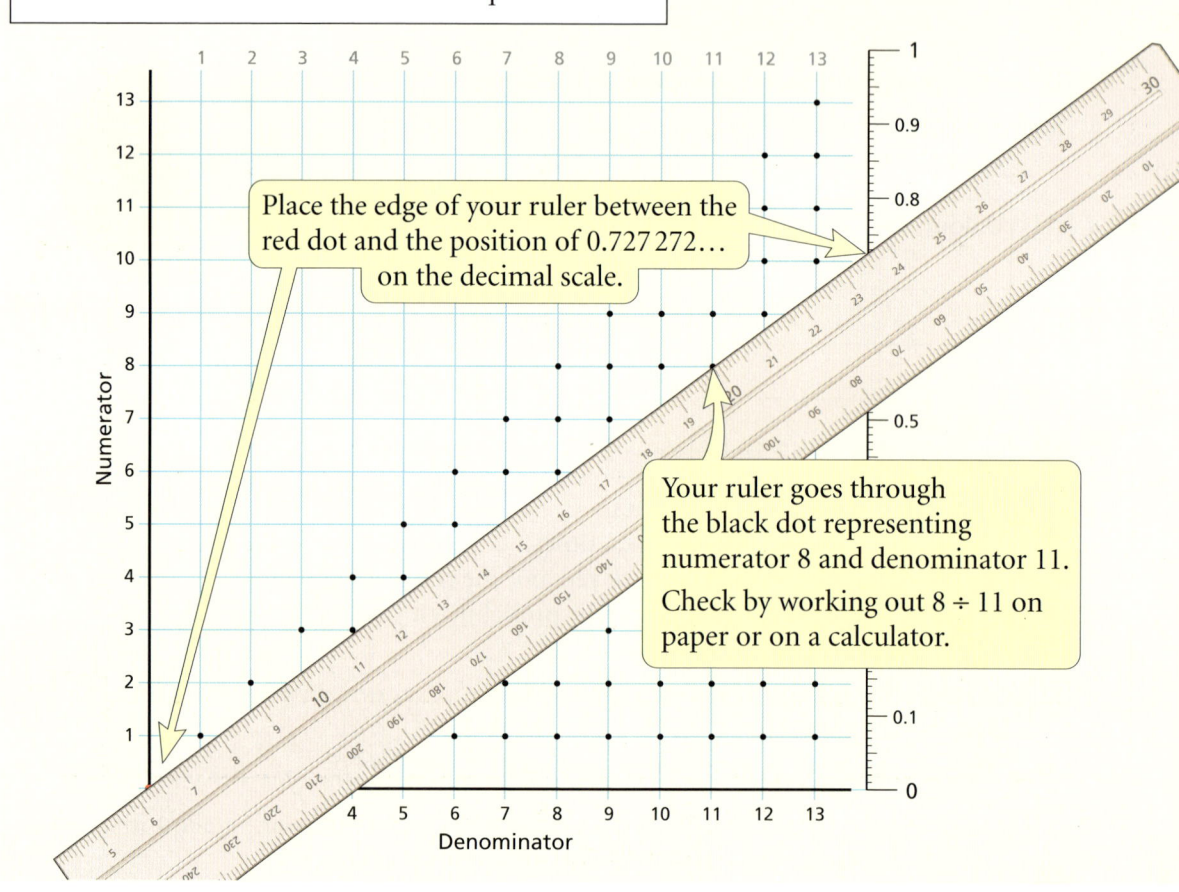

Place the edge of your ruler between the red dot and the position of $0.727\,272\ldots$ on the decimal scale.

Your ruler goes through the black dot representing numerator 8 and denominator 11. Check by working out $8 \div 11$ on paper or on a calculator.

E1 Use the chart to find an exact fraction for each of these. Check you have got the right dot by dividing.

(a) $0.166\,66\ldots$
(b) $0.4444\ldots$
(c) $0.636\,363\ldots$
(d) $0.583\,333\ldots$
(e) $0.615\,384\,615\,384\ldots$
(f) $0.428\,571\,428\,571\ldots$

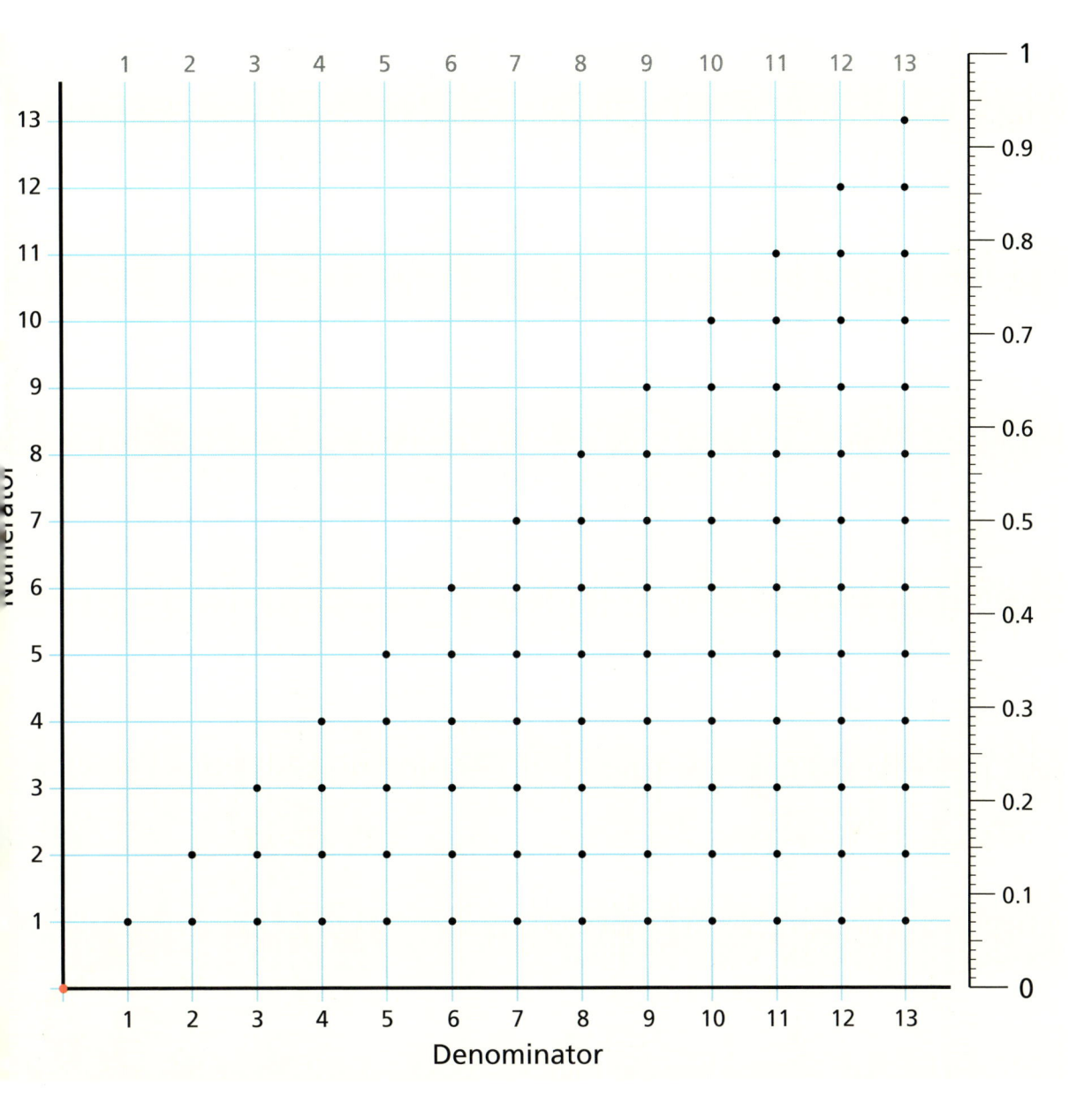

There is a 'shorthand' way of writing recurring decimals.

If only one digit recurs, it is shown once with a dot over it.

So $0.\dot{3}$ means $0.3333\ldots$ and $0.1\dot{6}$ means $0.16666\ldots$

If a group of digits recurs a dot is placed over the first and last digits in the group.

So $0.\dot{1}42\,85\dot{7}$ means $0.142\,857\,142\,857\,142\,857\ldots$

E2 Use the chart to find an exact fraction for each of these. Check by dividing.

(a) $0.\dot{6}$ (b) $0.8\dot{3}$ (c) $0.91\dot{6}$ (d) $0.\dot{9}\dot{0}$ (e) $0.\dot{8}57\,14\dot{2}$ (f) $0.\dot{3}07\,69\dot{2}$

What progress have you made?

Statement

I can use mental and written methods to calculate with decimals.

Evidence

1. Work these out in your head.
   (a) 7.3 + 11.8   (b) 43.8 − 2.9   (c) 1.8 × 4

2. Work out each missing number in your head.
   (a) 2.24 + ■ = 3.45   (b) 5.93 − ■ = 2.21
   (c) 2.7 × ■ = 10.8   (d) ■ × 5 = 14.0

3. Work these out without a calculator.
   (a) 0.85 × 100   (b) 50.1 ÷ 100   (c) 9.3 × 0.08
   (d) 66.6 ÷ 0.9   (e) 241 × 0.35   (f) 33 ÷ 0.015

I can solve problems involving decimals.

4.
   Apples £0.95 per kilo
   Grapes £1.90 per kilo
   Bananas £1.08 per kilo
   Pears £1.28 per kilo

   Work these out.
   (a) The cost of 0.4 kg of grapes
   (b) The cost of 1.2 kg of apples
   (c) The cost of 2.25 kg of bananas
   (d) The change from £10 when you buy 1.6 kg of apples and 1.5 kg of pears

5. A recipe for leaf green paint says 'mix yellow and blue in the ratio 5 : 3'.
   (a) If you use 6.5 litres of yellow, how much blue is needed?
   (b) How much of each colour is needed to make 12 litres of leaf green paint?

6. Anna has 82 copies of her favourite magazine. Each copy is 0.8 cm thick. She has some magazine files. Each file will take a maximum thickness of 9.4 cm. How many files does she need to store all her magazines?

18 Area of a circle

This work will help you
- calculate the area of a circle from its diameter or radius
- solve problems that involve circumference or area

A The formula for the area of a circle

- Explain why diagram A shows that the area of the circle is less than 4 times r^2.

- Explain why diagram B shows that the area of the circle is more than 2 times r^2.

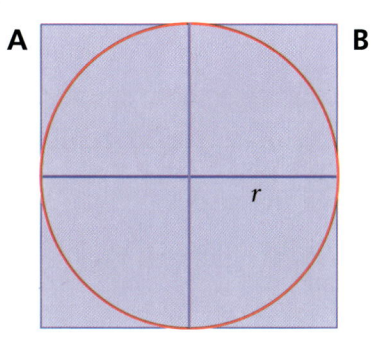

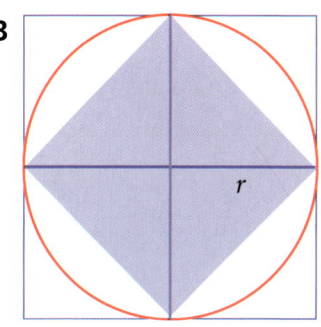

The areas of the circles below have been measured approximately by counting squares.

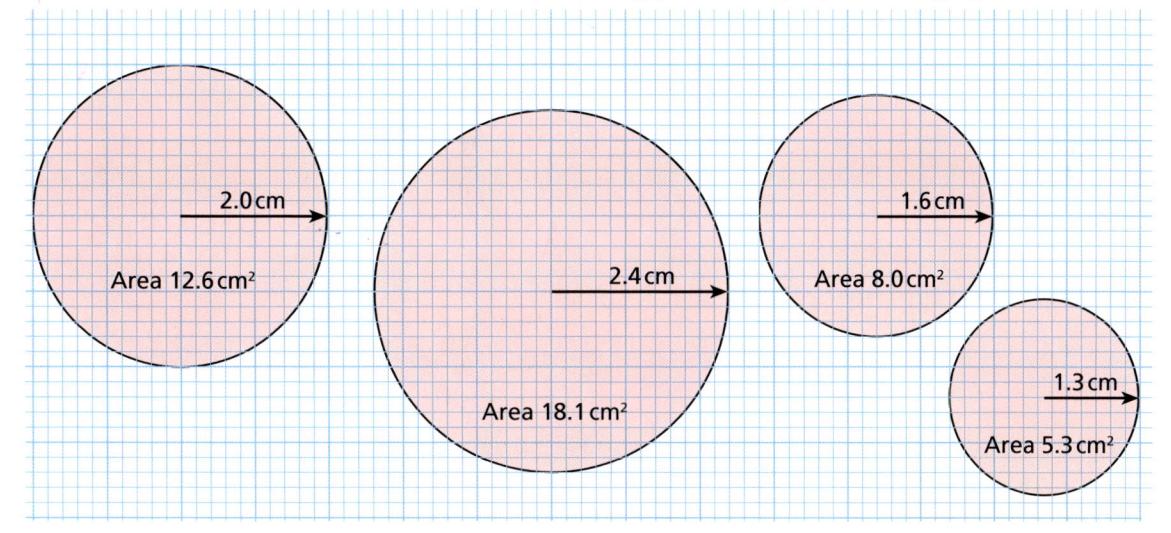

- 2.0 cm, Area 12.6 cm²
- 2.4 cm, Area 18.1 cm²
- 1.6 cm, Area 8.0 cm²
- 1.3 cm, Area 5.3 cm²

- Make a table showing r^2 and the area A for each circle. Work out how many times r^2 goes into A.

  What do you think is the formula connecting A and r?

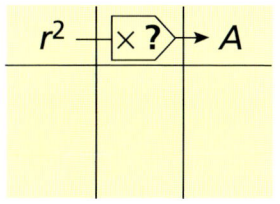

  $r^2 \ \boxed{\times \ ?} \ \rightarrow A$

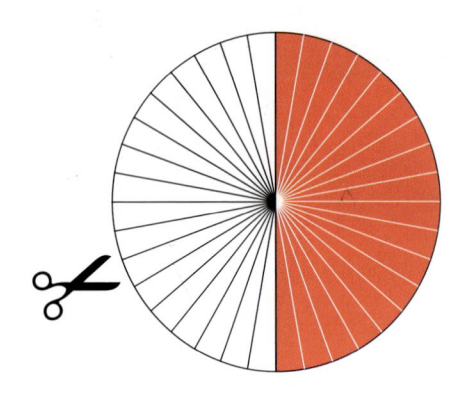

- Explain why these diagrams show that the formula for the area A of a circle of radius r is
$$A = \pi r^2.$$

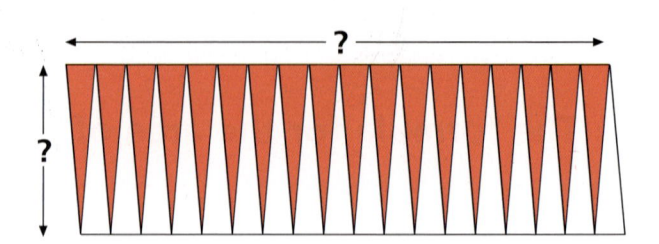

A1 Calculate, to the nearest $0.1\,\text{cm}^2$, the area of a circle whose radius is

(a) 6 cm (b) 7.5 cm (c) 2.4 cm (d) 5.6 cm (e) 0.9 cm

A2 (a) What is the radius of this circle?

(b) Calculate the area of the circle, to the nearest $0.1\,\text{cm}^2$.

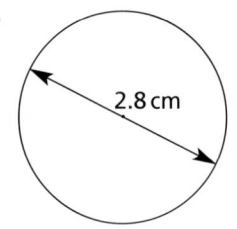

A3 Calculate, to the nearest $0.1\,\text{cm}^2$, the area of a circle with

(a) radius 4.4 cm (b) diameter 7.6 cm (c) radius 8.5 cm

(d) radius 2.7 cm (e) diameter 9.0 cm (f) diameter 5.3 cm

A4 Calculate the area of

(a) the outer circle, of radius 1.6 cm

(b) the inner circle, of radius 1.2 cm

(c) the purple space between the two circles

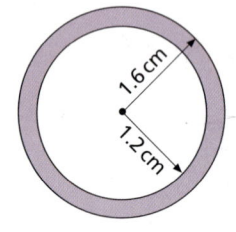

A5 Look at the diagram on the left.

(a) Do you think the area of the light green ring is greater than, equal to, or less than the area of the darker green circle?

(b) Find out whether you are right.

A6 Some men on board a ship are marking out a landing pad for helicopters.
It is a white circle of radius 4.5 m.
What is the area of the landing pad?

A7 Suppose you have an A4 sheet of paper (29.7 cm by 21.0 cm).
(a) What is the area of the largest circle you could cut out of the sheet?
(b) What area of the sheet is left over?

A8 Put these shapes in order of area size, smallest first.
Estimate by eye first, then check by measuring and calculation.

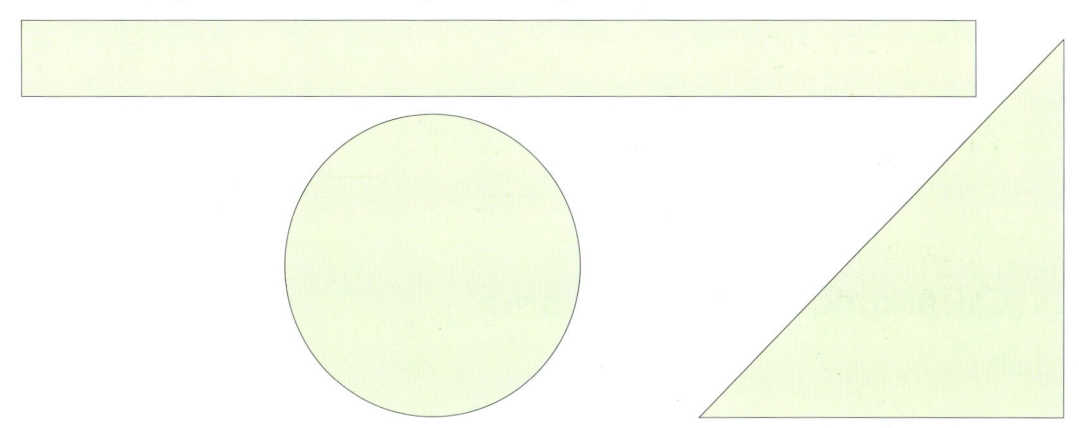

A9 Which is better value, a 20 cm diameter pizza costing £3
or two 10 cm diameter pizzas costing £2.50 altogether?

Give reasons for your answer.

B Area and circumference

Give all answers to one decimal place.

B1 Calculate (i) the circumference, (ii) the area of a circle of radius
(a) 2.2 cm   (b) 1.6 cm   (c) 5.1 cm   (d) 4.3 cm   (e) 16.2 cm

B2 (a) Calculate the area of a circle of radius 6.7 cm.
(b) Calculate the circumference of a circle of radius 9.7 cm.
(c) Calculate the area of a circle of **diameter** 24.3 cm.

B3 The curves at the end of a running track are semicircles.
(a) Mary runs round the outside of the track.
In the same time John runs round the inside.
How much further does Mary run?
(b) Calculate the area of the track (coloured).

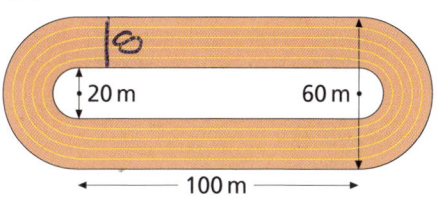

B4 Six coins all the same size are packed tightly into a rectangular tray as shown.
The diameter of each coin is 3 cm.

(a) How much space (shown pink) is wasted?

(b) What percentage of the tray is wasted?

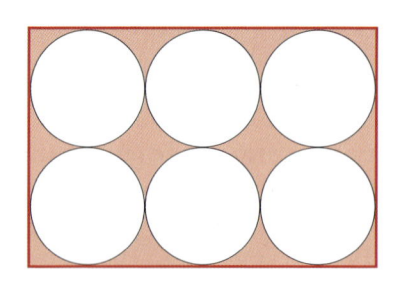

B5 A square picture is centred in a circular frame.
The diameter of the frame is 20 cm.
The corners of the picture are 5 cm from the edge of the frame.
Calculate the area of the frame (shown purple).

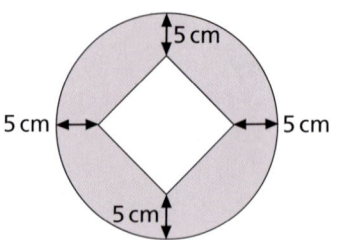

B6 Convert the lengths given in B1 to millimetres.
Calculate the areas of the circles in mm², to the nearest 10 mm².
Compare them with the answers in cm². What do you notice?

C Calculating radius given area

This is the flow diagram for the formula $A = \pi r^2$.

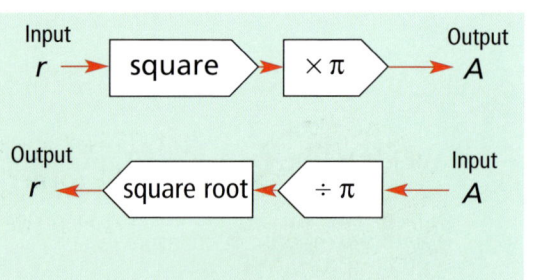

Reversing the diagram leads to the formula

$$r = \sqrt{\frac{A}{\pi}}$$

C1 What is the radius of a circle with area 12 cm²? Show your working.

C2 Find the radius of a circle with area

(a) 20 cm² (b) 50 cm² (c) 120 cm² (d) 3 cm² (e) 0.9 cm² (f) 1.5 cm²

C3 The area of a circle is 35 cm².

(a) Calculate the radius, leaving the result in the calculator display.

(b) Use the result to calculate the circumference of the circle, to the nearest 0.1 cm.

(c) Why should you **not** round the radius before using it to calculate the circumference?
(Try doing the calculations again, but this time rounding the radius.)

C4 Calculate, to the nearest 0.1 cm, the circumference of a circle of area 43.5 cm².

C5 Copy this table and fill in the missing values.

Radius	Diameter	Circumference	Area
12.8 cm			
	55.8 cm		
		72.6 cm	
			64.2 cm²

D Designs

D1 These designs are all made from parts of circles with radius 2 cm.
Calculate the blue areas.

(a) 4 cm

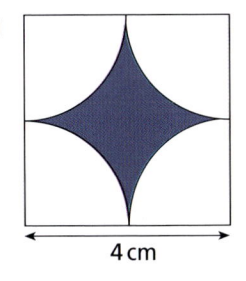

(b) 4 cm

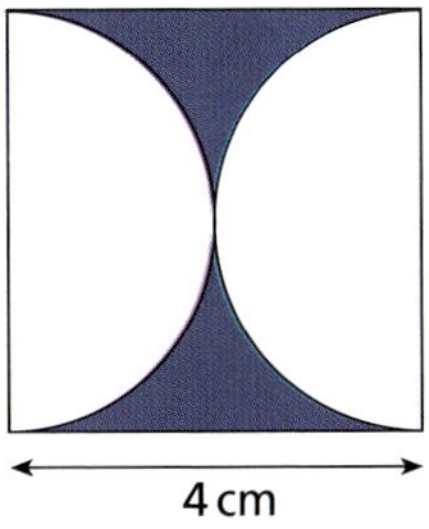

(c) 2 cm

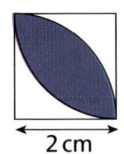

D2 In both of these designs the large circle has radius 2 cm.

The white circles touch at its centre.

The small green circles touch at the centres of the white circles.

What is the green area in each design?

(a) 4 cm   (b) 4 cm

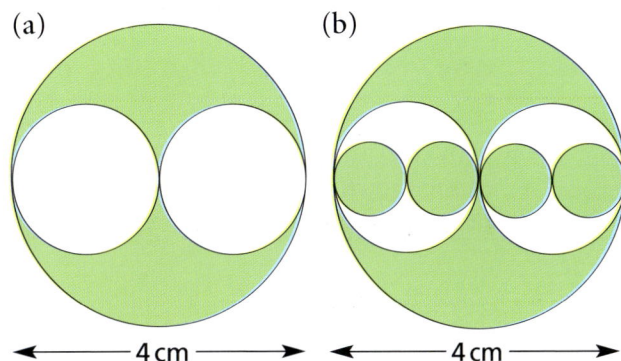

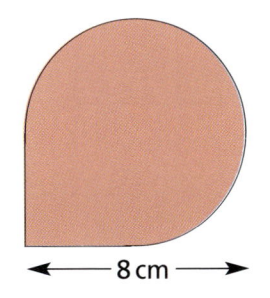

D3 This shape is made from a square and a circle with radius 4 cm.
The corner of the square is at the centre of the circle.

What is the area of the shape?

8 cm

What progress have you made?

Statement

I can find the area and the circumference of a circle.

I can find the area of a shape that involves circles or parts of circles.

Evidence

1 Find (i) the area, (ii) the circumference of a circle of radius
  (a) 5 cm   (b) 1.7 m   (c) 52 cm

2 What is the shaded area?

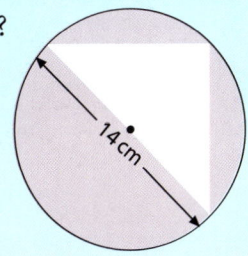

14 cm

131

Review 3

Do not use a calculator for questions 1 to 17.

1 Calculate $6.2 + 1.86 + 0.5$.

2 Write down how many metres are in 3.21 km.

3 The nth term of a sequence is $5n - 1$.
Write down the first five terms of this sequence.

4 Find the missing angle in this polygon.

5 Why does this chart give a misleading picture of the growth in the membership of a sports club?

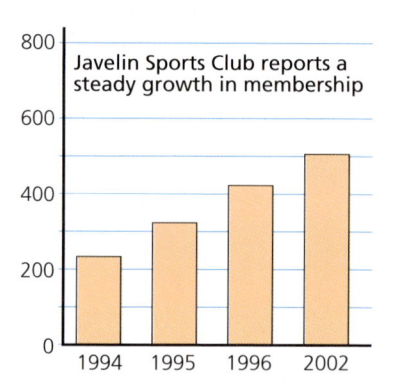

Javelin Sports Club reports a steady growth in membership

6 Decide which pair of numbers from the loop make this correct.

■ × ■ = 0.2

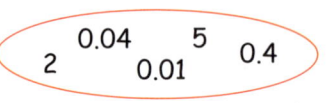

(0.04, 2, 0.01, 5, 0.4)

7 (a) Find an expression for the nth term of the linear sequence
2, 6, 10, 14, 18, …

(b) Use your expression to calculate the 20th term of this sequence.

8 Find the missing number in each of these.

(a) $5.8 + ■ = 7.35$
(b) $■ × 6 = 19.14$
(c) $3.06 - ■ = 2.52$

9 The yellow shape is a regular polygon.
(a) What is the size of each exterior angle?
(b) What is the size of each interior angle?

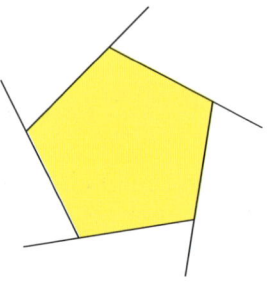

10 Decide whether each of these is true or false.

(a) $(0.4)^2 = 0.8$ (b) $(0.3)^2 = 0.9$ (c) $\sqrt{0.01} = 0.1$ (d) $(0.2)^3 = 0.8$

11 Calculate 5.81×19.

12 Work out the volume and surface area of this cuboid.

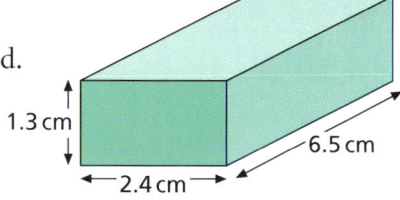

1.3 cm
2.4 cm
6.5 cm

13 These are the ages of a sample of people using a library one Saturday afternoon.

| 40 | 65 | 14 | 47 | 61 | 69 | 59 | 51 | 5 | 71 | 6 | 31 | 54 | 22 | 63 | 59 | 13 | 68 | 51 | 60 | 58 |
| 39 | 6 | 45 | 62 | 69 | 36 | 14 | 56 | 42 | 8 | 15 | 28 | 41 | 10 | 10 | 74 | 48 | 11 | 60 | 63 | 53 |

(a) Draw a stem and leaf diagram for the data.

(b) Describe the shape of the distribution.
Suggest some hypotheses that might account for its shape and which could be investigated further.

(c) Find the median age from the diagram.
Is it a useful measure in this case?

14 Work these out.

(a) $\dfrac{9}{0.3}$ (b) $\dfrac{6.3}{0.7}$ (c) $\dfrac{4.42}{1.3}$ (d) $7.8 \div 1.2$

15 Match up each fraction with a decimal.

$\dfrac{1}{4}$ $\dfrac{2}{3}$ $\dfrac{5}{100}$ $\dfrac{2}{9}$ $\dfrac{4}{5}$ 0.8   0.05   0.222 22...   0.25   0.666 66...

16 This is part of a regular polygon. How many sides does it have?

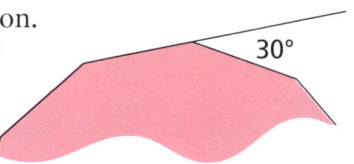

30°

17 Find the *n*th term of the linear sequence: 37, 34, 31, 28, 25, ...

18 These are the inside measurements of five picture frames.

Frame	A	B	C	D	E
Width (cm)	22.5	15.0	14.0	16.5	20.5
Height (cm)	27.0	20.0	20.9	24.6	28.9

(a) Calculate the ratio $\dfrac{\text{height}}{\text{width}}$ for each frame, to 2 d.p.

(b) Which of the frames would fit an enlargement made from a photo 10.2 cm wide by 15.2 high?

19 The area (A cm^2) of some circles with various radii (r cm) are shown in the table. The area is rounded to 2 d.p.

r	1	2	3	4
A	3.14	12.57		

(a) Copy and complete the table.

(b) Work out the ratio $\frac{A}{r}$ (correct to 2 d.p.) for each pair of values.

(c) Explain how your ratios show that A is not directly proportional to r.

20 This table shows the number of members in a table tennis club at the end of each of the years shown.

Year	1997	1998	1999	2000	2001	2002	2003
Boys	7	7	7	5	6	7	8
Girls	4	5	5	4	5	8	9

(a) On one grid, draw time series graphs for this data.

(b) Describe how the membership changed over these years.

(c) What percentage of the members were girls (i) in 1997   (ii) in 2003

21 A brooch is made from a silver disc with a smaller disc punched out.

The silver disc had a diameter of 3.6 cm.
The circular hole has a diameter of 1.4 cm.

One side of the brooch is then polished.
Calculate the area that is polished.

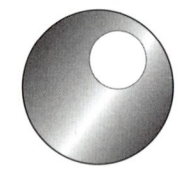

22 In the table, b is directly proportional to a.
Find the value of n.

a	1.6	2.8	3.4
b	4.0	n	8.5

23 These are the radii in centimetres of five circles.

      1.4   2.6   4.3   2.8   0.9

(a) Find the mean and median of these radii.
Which is greater, mean or median?

(b) Calculate the circumferences of these circles to 1 d.p.

(c) Find the mean and median of the the circumferences.
Which is greater?
By what scale factor is the mean of the circumferences greater than the mean of the radii?

(d) Calculate the areas of the circles to 1 d.p.

(e) Find the mean and median of the areas. Which is greater?

24 A circular pond has an area of 300 m^2.
Find the radius of this pond, correct to 2 decimal places.

19 The right connections

This is about looking for relationships between two sets of data.
The work will help you
- use and draw scatter diagrams and lines of best fit
- understand types of correlation

A 'The Missing Link'

Lemon   Biffo   Aaron   Phil   Howie   Stix

The picture above shows the rock group 'The Missing Link' on the cover of their latest album 'New Connections'.

The **scatter diagram** on the right shows their heights and hair lengths.

- Can you work out which point represents each member of the group?

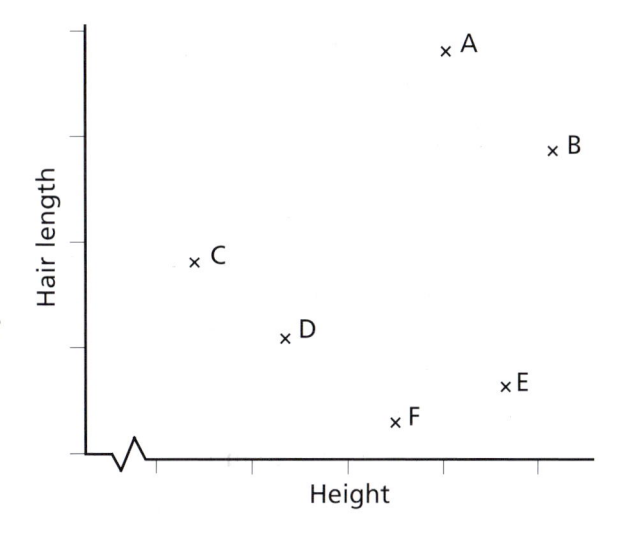

B Drawing scatter diagrams

These are some common British moths.
Is there a connection between body length and wingspan?

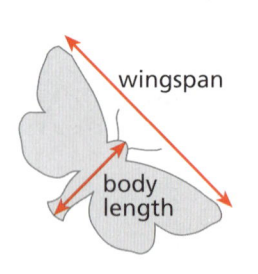

The names of these moths
A Goat
B Emperor
C Garden Tiger
D Lime Hawk
E Cream Spot Tiger
F Puss
G Peppered
H Wood Tiger
I Yellow Underwing
J Drinker
K Dot
L Scarlet Tiger
M Coxcomb Prominent
N Hook Tip
O Six Spot Burnet
P Ghost
Q Hornet Clearwing
R Jersey Tiger

The body length and wingspan of moths A and B have been plotted on this scatter diagram.

B1 Copy the scatter diagram and plot points on it for all the other moths.

(a) Does the moth with the longest body have the largest wingspan?

(b) Does the moth with the shortest body have the smallest wingspan?

(c) Is it generally true that the moths with longer bodies have larger wingspans?

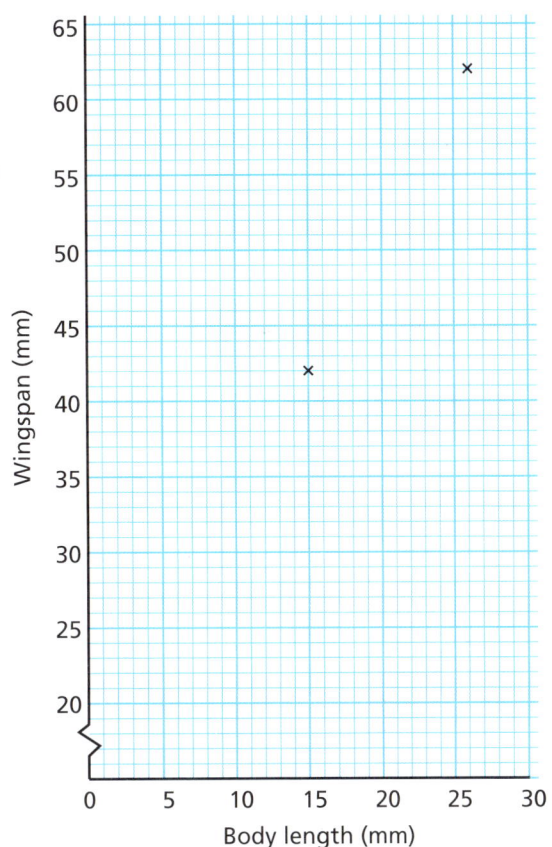

B2 A group of pupils had their pulse rate at rest measured (in beats per minute).
Then they did a 'bleep test', which involves fast running. A high bleep score means a good performance. After the test, the 'recovery time' needed for their pulse to return to normal was recorded.

(a) Draw a scatter diagram for pulse at rest and bleep score.

Start your axes like this.

What does your diagram suggest about bleep score and pulse at rest?

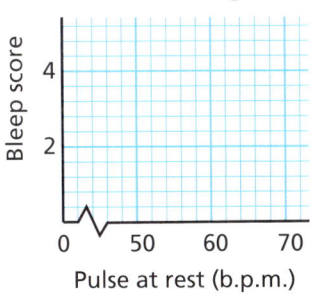

(b) Draw a scatter diagram for pulse at rest and recovery time. What does your diagram suggest about recovery time and pulse at rest?

(c) Draw a scatter diagram for recovery time and bleep score. What does it suggest?

Pulse at rest (b.p.m.)	Bleep score	Recovery time (min)
60	8.5	4.0
54	11.2	3.0
67	12.0	5.0
69	8.8	3.0
72	9.5	3.0
76	9.3	3.5
59	9.3	2.0
65	10.0	4.0
68	9.4	3.0
90	6.8	5.0
79	5.8	4.5
88	5.3	4.5
80	7.0	6.0
96	3.6	5.5
82	6.5	4.0
92	3.7	3.5
76	6.0	4.0
83	4.3	4.5
86	6.2	5.0
70	7.3	4.5

C Correlation and lines of best fit

C1 What does each scatter diagram below show?
Write a sentence about each one, for example 'Taller people tend to be heavier'.

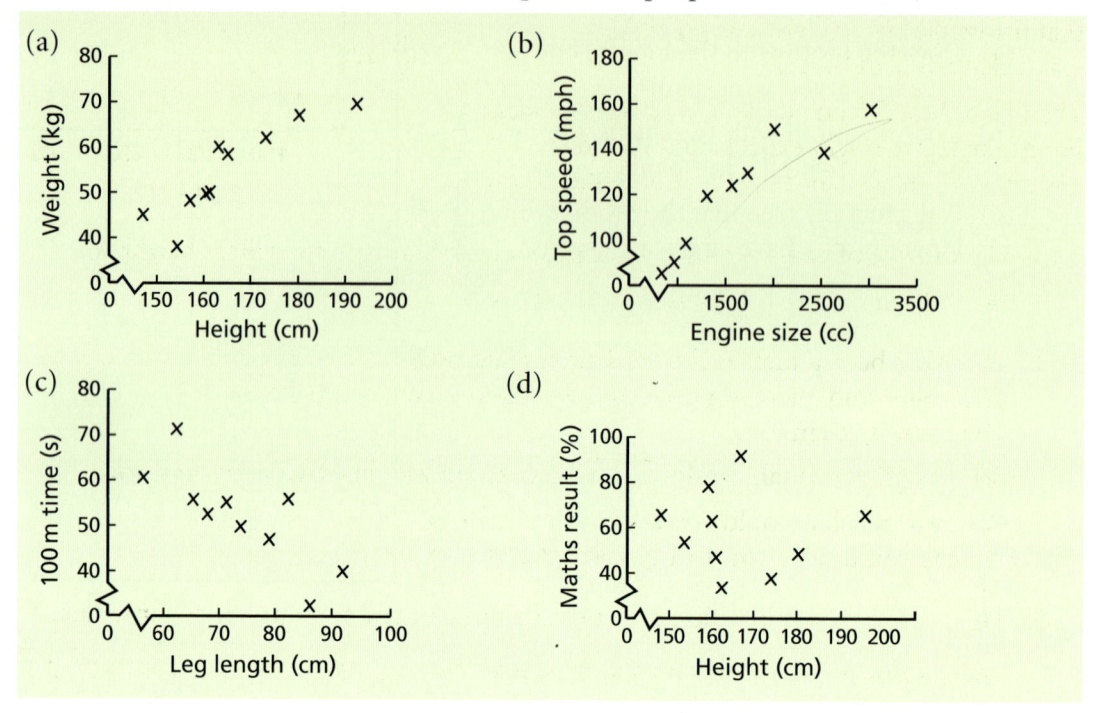

(a) Weight (kg) vs Height (cm)

(b) Top speed (mph) vs Engine size (cc)

(c) 100 m time (s) vs Leg length (cm)

(d) Maths result (%) vs Height (cm)

C2 Sketch the scatter diagrams that you think these sets of data would give you.

(a) The time spent by pupils watching TV and and the time spent on homework

(b) Hand span and hair length

The scatter diagrams you have seen are of these three types.

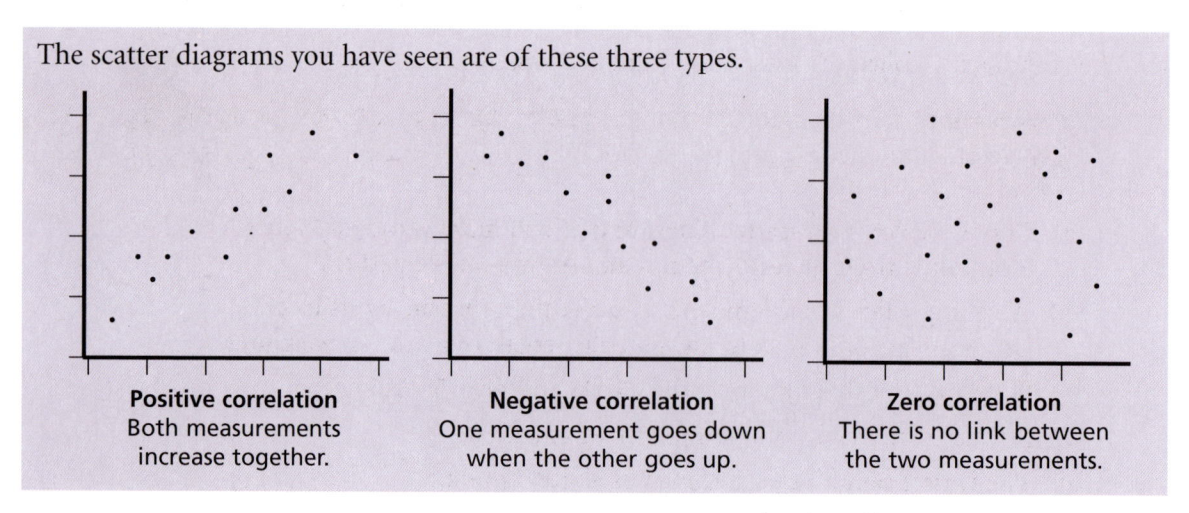

Positive correlation
Both measurements increase together.

Negative correlation
One measurement goes down when the other goes up.

Zero correlation
There is no link between the two measurements.

C3 Which of these types is each diagram in C1?

This scatter diagram shows some pupils' heights and foot lengths.

We can see there is correlation between the two measurements, so we can draw a straight line that fits the points as closely as possible.
It is called a **line of best fit**.

The line of best fit can be used to make estimates. For example, we might expect a pupil who is 160 cm tall to have a foot length of about 21.5 cm.

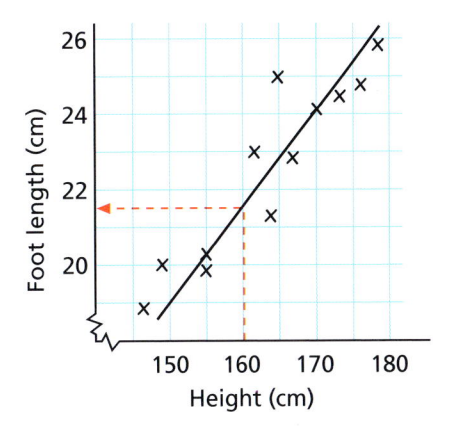

C4 Use the line of best fit to estimate the foot length of a person whose height is

(a) 150 cm (b) 165 cm (c) 175 cm

C5 There has been a theft at the local swimming pool.
The police find a footprint which they believe to be the thief's.
The print is 25 cm long.

(a) Roughly how tall would you say the person was who made this footprint?

(b) How reliable would you say your estimate is?
Do you think a court would accept this as evidence?

C6 This table gives information about the engine size and horsepower of some cars.

Engine size (litres)	1.8	2.8	3.5	2.9	1.6	1.4	1.3	1.3	3.2	1.7	1.5	1.9	2.0	1.2
Horsepower	59	193	202	191	100	75	62	74	216	130	75	140	115	44

On graph paper, choose scales for a scatter diagram that will make it a reasonable size and roughly square.

(a) Draw a scatter diagram and line of best fit.

(b) Use the line to estimate the horsepower of a car whose engine size is 2.5 litres.

C7 A fish farmer collected this information about salmon.

Weight (kg)	5.0	3.3	4.3	5.1	3.4	5.1	4.7	4.2	3.1	2.9	4.5	4.9	3.4	5.0	5.0
Length (cm)	74	49	62	67	54	74	71	57	54	48	69	69	58	79	71

(a) Choose scales for a scatter diagram that will make it roughly square.
Plot points from the information and draw a line of best fit.

(b) Weighing a live salmon in a net is easier than measuring its length.
Use your line of best fit to estimate the length of a salmon weighing

(i) 4 kg (ii) 3 kg (iii) 3.7 kg (iv) 5.5 kg

(c) Estimate the weight of a salmon of length 75 cm.

(d) The largest salmon ever caught was about 1 m long.
Use your line of best fit to estimate its weight.
How reliable do you think this estimate is?

C8 Alex is investigating the price of used cars.
She finds some adverts for the same make and model of car but of different ages in her local paper.

Age (years)	0	2	2	3	4	4	5	5	8	9	10	12
Price (£)	8300	7300	7250	7000	4700	5400	3700	3800	2400	2500	1400	1000

(a) Plot this information on a scatter diagram and draw a line of best fit.

(b) Use your graph to estimate the price of a 7-year-old car.

(c) After how many years roughly should the car have zero value?

D Quarters

Sometimes it is hard to decide whether there is any correlation between two sets of measurements.

This scatter diagram shows some people's heights and footlengths.

The median height (found by putting the heights in order) is 160 cm. The median footlength is 20.9 cm. The medians are shown on the diagram.

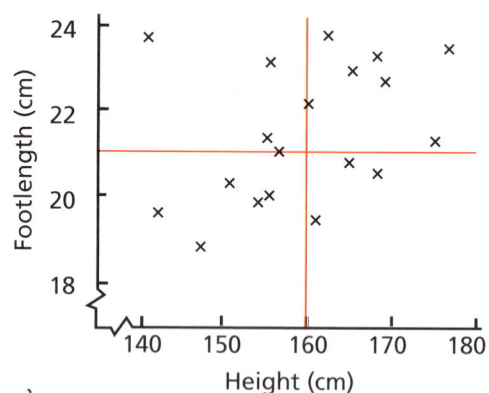

The points in each pair of diagonally opposite 'quarters' can be counted, ignoring those on a line.

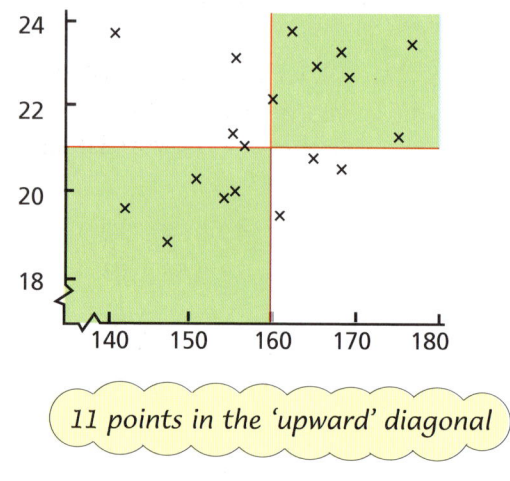

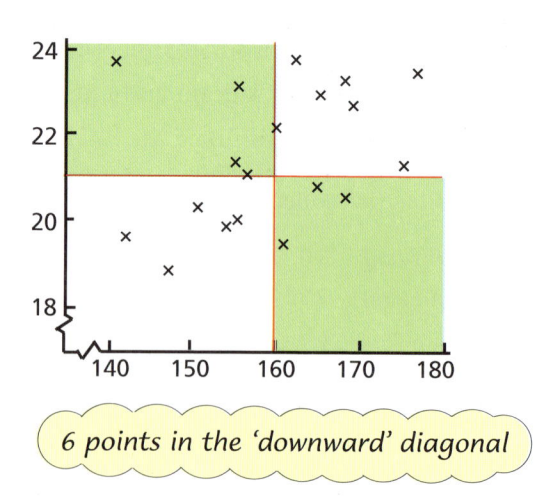

11 points in the 'upward' diagonal

6 points in the 'downward' diagonal

There are many more points in the 'upward' (↗) diagonal than in the 'downward' (↘). This indicates a positive correlation.

D1 What would happen if there were

(a) negative correlation

(b) zero correlation

D2 This table gives information about 25 cars.
It shows the mass of carbon monoxide (CO) and of
nitrogen dioxide (NO_2) given off for each kilometre driven.

(a) Draw a scatter diagram from the data.

(b) Find the median CO emission and the median NO_2 emission. Use them to draw lines that make 'quarters' on your scatter diagram.

(c) Use the quarters to see if there is any correlation between CO and NO_2 emissions.

CO (g/km)	NO_2 (g/km)
3.1	0.8
9.1	0.4
5.3	0.7
2.7	0.8
3.1	0.7
4.5	0.9
4.7	0.7
7.6	0.8
9.1	0.4
4.8	0.8
7.2	0.8
2.5	0.9
4.6	0.9
3.2	0.7
2.5	1.2
3.7	0.6
1.3	1.1
3.9	0.9
9.2	0.7
2.9	1.1
6.0	0.4
9.3	0.3
11.8	0.5
2.6	0.7
1.4	1.2

Avoiding nonsense

A teacher has a maths test result and a French test result for each pupil in a group of 30.
She draws a scatter graph to show the data.

After she has plotted five points the maths and French results seem to have negative correlation.

When ten points have been plotted there seems to be zero correlation.

When points for the whole class have been plotted there is weak positive correlation.

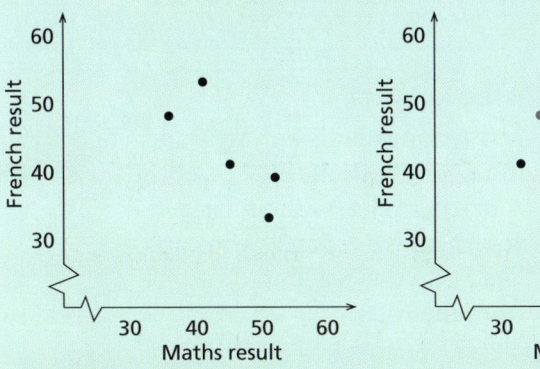

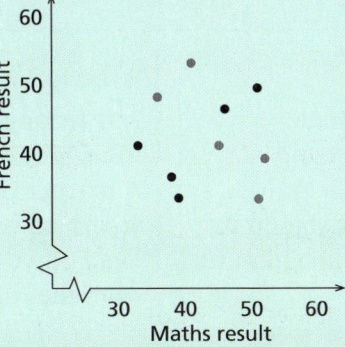

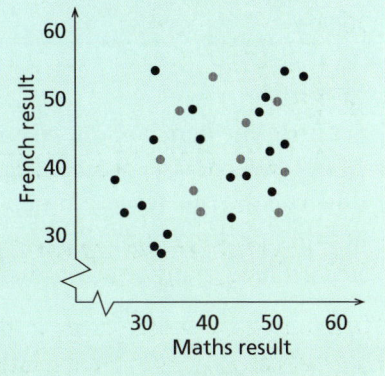

Often a small set of data seems to show a type of correlation, but adding more data points leads to a different conclusion.
Apparent correlation in small data sets is sometimes called 'nonsense correlation'.

E Correlation in surveys and experiments

Drawing a scatter diagram and looking for any correlation is one way to explore whether two measurements are related.

These are things to consider.

- What hypothesis do you want to test?
- Where will you collect data from?
- What sort of data collection sheet will you use?
- How much data might you need to be reasonably sure you do not get 'nonsense correlation'?
- Could you investigate correlation using data that somebody else has collected and arranged in a table?

The fitness tests below can provide data to plot on a scatter graph and examine for correlation.

Your hypothesis might be 'People with long legs do better in a shuttle run', so you would need to record leg length. Or you could record height, or anything else you think might be related to performance in the tests.

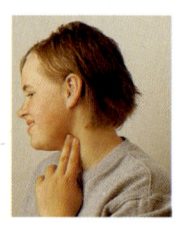

Before you start any of the tests, sit down for at least two minutes and then take your pulse (in beats per minute). It is easiest to do this at the side of your throat.
'Pulse at rest' can then be one of the measures you use in your investigation.

Shuttle run

This is a good test of speed and mobility.
The course is made with cones placed 4.5 m apart.
Time how long it takes to complete the run.

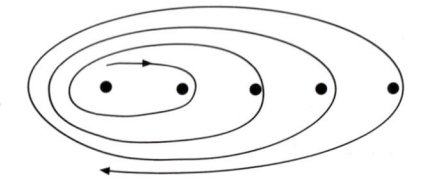

Step-ups

Set up a safe bench about 40 cm high.
A 'step-up' consists of stepping up and down again, one leg at a time. It helps if another person counts 1-2-3-4 as you do it.
Record how many you complete in a minute.

Reaction time

A reaction ruler is an easy way to record reaction time.
A partner releases the ruler and you catch it as soon as you can.
Read off your reaction time from the ruler.

Holding power

Hold a school bag at arm's length.
Time how long you can do this for.

What progress have you made?

Statement

I can draw scatter diagrams from data.

Evidence

1. This table shows results recorded by some students in an after school athletics club. Draw a scatter diagram for the data.

Height (cm)	Jump (m)
163	1.3
168	1.4
147	1.1
140	1.1
192	1.9
156	1.2
180	1.7
177	1.5
158	1.4
160	1.2
174	1.6
172	1.4

I can describe connections between measurements.

2. Describe the connection between height and distance jumped.

I can tell what type of correlation the data shows.

3. (a) What type of correlation does this data show?

   (b) What other types of correlation are there? Sketch a scatter diagram to illustrate each one.

I can draw a line of best fit and use it to estimate values.

4. (a) Draw a line of best fit.

   (b) A student 185 cm tall transfered from a similar club. How far would you expect this student to be able to jump?

   (c) Three students were absent when the club recorded the results above, but they had a go later, getting these results.

	Height (cm)	Jump (m)
Sam	155	1.1
Mel	166	1.6
Chris	180	1.6

   Say whether each one did about what you would expect, or did better or did worse.

20 Algebra problems

This work will help you
- form and solve equations with subtracted unknowns

A Take off

A1 Solve these equations. Show all your working.
Check your answers are correct.

(a) $4x = x + 30$
(b) $7x + 8 = 5x + 32$
(c) $7x - 16 = 4x + 23$
(d) $x - 2 = 5x - 22$
(e) $2(a + 11) = 5a - 17$
(f) $3(t - 2) = 4t - 10$
(g) $3(y - 5) = 5(y - 7)$
(h) $6u - 4 = 4(u + 1)$

A2 Rhys thinks of a number.
He multiplies the number by 7.
Then he takes off 25.
The answer is 5 more than the number he first thought of.

Turn his number puzzle into an equation and solve it to find the number he first thought of.

A3 Geraint has 3 bags of coins and 36 extra coins.

Ciara has 5 bags of coins and 8 extra coins.

Suppose that there are n coins in each bag.

(a) Write an expression for the number of coins Geraint has.
(b) Write an expression for the number of Ciara's coins.
(c) They each have the same number of coins.
Write an equation and solve it to find how many coins there are in a bag.

*A4 Imran thinks of a number.
He multiplies his number by 7 and takes the result away from 120.
He gets the same answer if he adds 16 to his original number.

What number was Imran thinking of?

B Subtracted unknowns

$$120 - 7n = n + 16$$ (add 7n to both sides)
$$120 = 8n + 16$$ (take 16 off both sides)
$$104 = 8n$$ (divide both sides by 8)
$$13 = n$$

$$15 - 4x = 60 - 7x$$ (add 7x to both sides)
$$15 + 3x = 60$$ (take 15 off both sides)
$$3x = 45$$ (divide both sides by 3)
$$x = 15$$

B1 (a) Copy and complete this working to solve $36 + 3x = 100 - 5x$.

(b) Check that your answer works in the original equation.

$$36 + 3x = 100 - 5x$$ (add 5x to both sides)
$$36 + \ldots = 100$$ (take 36 off both sides)
$$\ldots = 64$$ (divide both sides by 8)
$$\ldots = \ldots$$

B2

$$25 - x = 40 - 4x$$ (add 4x to both sides)
$$\ldots = \ldots$$ (take 25 off both sides)
$$\ldots = \ldots$$ (divide both sides by 3)
$$\ldots = \ldots$$

(a) Copy and complete this working to solve $25 - x = 40 - 4x$.

(b) Check that your answer works in the original equation.

B3 Here is Goldie's homework. She has not done very well! Write out correct answers for Goldie.

(a) $25 - x = 10 + 2x$ (take x from both sides)
$25 = 10 + x$ (take 10 from both sides)
$15 = x$ ✗

(b) $4n - 15 = 5 - n$ (take 4n from both sides)
$15 = 5 - 5n$ (take 5 from both sides)
$10 = 5n$ (divide both sides by 5)
$2 = n$ ✗

(c) $11 - 2d = 35 - 4d$ (take 4d from both sides)
$11 - 6d = 35$ (take 11 from both sides)
$6d = 24$ (divide both sides by 6)
$d = 4$ ✗

(d) $6j - 20 = 100 - 2j$ (add 2j to both sides)
$8j - 20 = 100$ (take 20 from both sides)
$8j = 80$ (divide both sides by 8)
$j = 10$ ✗

B4 Solve each of these equations.

(a) $5f + 1 = 100 - 4f$
(b) $100 - 3a = 38 - a$
(c) $34 - 3w = 100 - 9w$
(d) $27 - 4e = 6 + 3e$
(e) $27 + 5g = 27 - 3g$
(f) $91 - 5x = 16$
(g) $26 - 9h = 21 - 7h$
(h) $35 - 2r = 3r + 5$
(i) $7y - 51 = 49 - 3y$
(j) $50 - h = h - 50$

C Number puzzles

Jordan:
I think of a number.
I multiply it by 3.
I take my result away from 96.
The answer is 66.
What number was I thinking of?

Abbie:
I think of a number.
I multiply it by 3.
I take my result away from 96.
My answer is 10 more than the number I started with.
What number was I thinking of?

C1 Here are two number puzzles and four equations.

Puzzle 1
I think of a number.
I multiply it by 4 and take the result away from 12.
My answer is twice my starting number.
What number did I start with?

Puzzle 2
I think of a number.
I multiply it by 4 and take away 12 from the result.
My answer is twice my starting number.
What number did I start with?

A $12 - 4n = 2$
B $4n - 12 = 2n$
C $12 - 4n = 2n$
D $4n - 12 = 2$

(a) Which puzzle fits which equation?
(b) Use the correct equation to solve each puzzle.

C2 Solve each of these number puzzles by writing an equation and solving it.

(a)
I think of a number.
I double it and take the result away from 25.
The answer is 3 times the number I started with.
What number did I start with?

(b)
I think of a number.
I multiply it by 3.
I take the result away from 63.
The answer is 7 more than the number I started with.
What number did I start with?

C3 Solve each of these number puzzles.

(a) I think of a number.
I multiply it by 6.
I take the result away from 48.
The answer is double my starting number.
What number did I start with?

(b) I think of a number.
I multiply it by 4.
I take the result away from 67.
I get the same answer if I multiply my starting number by 6 and take the result away from 83.
What number did I start with?

(c) I think of a number.
I multiply it by 5.
I take the result away from 33.
The answer is 3 less than the number I started with.
What number did I start with?

(d) I think of a number.
I multiply it by 5.
I take the result away from 62.
The answer is one less than double my starting number.
What number did I start with?

C4 David and Bianca are playing number puzzles.

David: I multiply my number by 5 and then take the result away from 36.

Bianca: I multiply my number by 2 and then take 6 away from the result.

They both start with the same number, and their answers are the same.
What number did they start with?

C5 David and Bianca start with another number.

David multiplies his number by 6, and then takes the result away from 86.
Bianca doubles her number and takes the result away from 46.

They both get the same answer.
What number did they both think of at the start?

C6 Copy and complete the solution of this equation.

$3(4 - y) = 2y - 3$ (multiply out brackets)
$12 - 3y = 2y - 3$ (add 3y to both sides)
$12 = \ldots - 3$ (add ... to both sides)
$\ldots = \ldots$
$\ldots = \ldots$ (divide both sides by ...)

C7 Solve each of these equations.

(a) $5(10 - j) = 3j + 2$
(b) $3(40 - t) = 2(20 + t)$
(c) $3 - 2f = 3(f - 4)$
(d) $h - 2 = 3(10 - h)$
*(e) $4(2 - d) = 3(3 - d)$
*(f) $2(5 - 3a) = 2 - 10a$

C8 Emma and Jake are playing number puzzles.
They start with the same number.

Emma doubles her number and subtracts the result from 12.
Jake subtracts his number from 6 and then doubles the result.

They both get the same answer.
What number did they both think of at the start?

*__C9__ What happens if you try to solve the equation $3(5 - x) = 10 - 3x$?
Can you explain this?

D Mixed problems

David has 75 pence.
He buys 6 toffee chews.

Sophie has 61 pence.
She buys 4 toffee chews.

They both end up with the same amount of money.
How much does a toffee chew cost?

*__D1__ Lauren and Rebecca are buying CDs.
Lauren starts the day with £43; Rebecca starts with £79.

Lauren buys 4 CDs and Rebecca buys 10. Each CD costs the same.
After buying the CDs they each have the same amount of money left.

How much does a CD cost?

*__D2__ Daniel and Joshua are making bows out of ribbon.
They each start with a long strip of ribbon
and cut short pieces off for each bow.

Daniel starts with a strip of ribbon 320 cm long and makes 17 bows.
Joshua starts with a strip 500 cm long and makes 27 bows.

They each have the same length of ribbon left over.

(a) How many centimetres of ribbon are there in one bow?

(b) How many centimetres of ribbon do they each have left over?

*D3 Sophie and Matthew each have some packets of Fruitees.

Sophie starts with 5 packets.   Matthew starts with 3 packets.

Sophie eats 62 Fruitees from her packets, and Matthew eats 28 from his.
Now they each have the same number of Fruitees left.

(a) How many Fruitees are in a single packet?

(b) How many Fruitees do they each have left?

*D4 Chloe and Callum are packing chocolates into boxes.
Each box holds the same number of chocolates.
Chloe has 789 chocolates to pack; Callum has 657.

When they stop for a tea-break, Chloe has
packed 27 full boxes and Callum has packed 21.

They notice that they now each have the same number
of chocolates left to pack.

(a) How many chocolates does a box hold?

(b) How many chocolates did they each have left at tea-break?

*D5 Look at this set of patterns. Each one is made from hexagonal tiles.

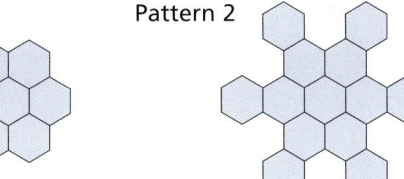

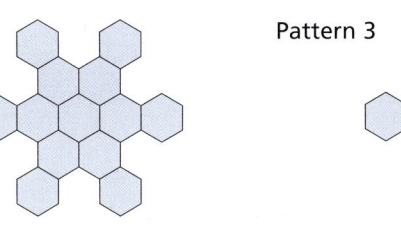

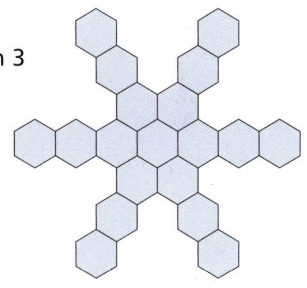

Pattern 1   Pattern 2   Pattern 3

(a) Make a table going up to pattern
number 6, showing how many
tiles there are in each pattern.

Pattern number	1	2	3	4	5	6
Number of tiles	7					

(b) Copy and complete this formula:

Number of tiles in the nth pattern = ...n + ...

(c) In one of these patterns, there are 1327 tiles!
Using your answer to (b), write down an equation for n.

Solve your equation to find the pattern number which has 1327 tiles.

Equation puzzles

Here are some equations to solve.
You may not have met ones like these before.
But you can still try to puzzle them out!

$a^2 = 25$	$b^2 + 64 = 100$	$2c^2 = 98$	$\dfrac{d^2}{2} = 32$
$\dfrac{8}{e} = 4$	$\dfrac{1}{f} = \dfrac{1}{2}$	$\dfrac{4}{g} = \dfrac{1}{4}$	$\dfrac{1}{h} = 10$
$\dfrac{i+5}{2} = 1$	$\dfrac{2j+1}{15} = 1$	$\dfrac{5k-3}{12} = 1$	$\dfrac{1}{2l-7} = 1$

Some of the puzzles have more than one answer. Did you find them?

What progress have you made?

Statement

I can solve equations where the unknowns are subtracted.

Evidence

1 Solve these equations. Show all your working and check that your answers are correct.

   (a) $4x + 5 = 54 - 3x$

   (b) $x - 2 = 40 - 5x$

   (c) $30 - 7x = 6 - 5x$

   (d) $100 - 8x = 20$

I can form equations from number puzzles where the unknowns are subtracted.

2 Write down and solve an equation from this number puzzle.

> I think of a number.
> I multiply it by 7.
> I take my result away from 99.
> My answer is 4 times the number I started with.
> What number was I thinking of?

3 Hannah and Lucy both think of the same number.

   Hannah multiplies her number by two and then adds 10.
   Lucy multiplies her number by three, and then takes the result away from 75.

   They both end up with the same number.
   What number did they both think of?

21 Angles

This work will help you
- revise earlier work on angles
- learn about angles and parallel lines

A Angle relationships – revision

A1 What do angles round a point add up to?

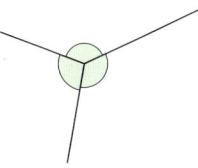

A2 What do angles on a straight line add up to?

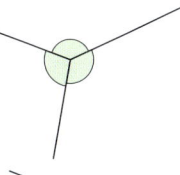

A3 What do the three angles of a triangle add up to?

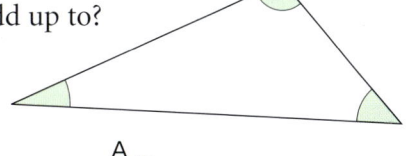

A4 This is an isosceles triangle (the marks mean the sides AB and BC are the same length). What can you say about the angles?

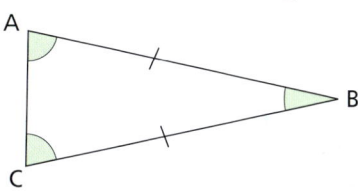

A5 These are vertically opposite angles (sometimes called opposite angles at a vertex). What can you say about them?

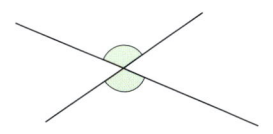

A6 This shape has rotation symmetry of order 3, but no reflection symmetry. Are any angles equal? If so, which?

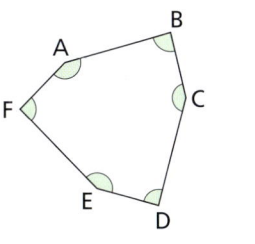

A7 This is a parallelogram.
   (a) Describe any symmetry it has.
   (b) Are any angles equal? If so, which?

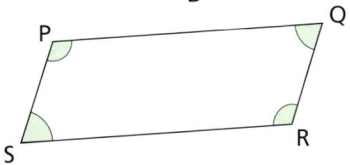

Get your answers for A1 to A7 checked before you go any further.

A8 For each of these, find the missing angle and say which of these angle relationships you used to get your answer.

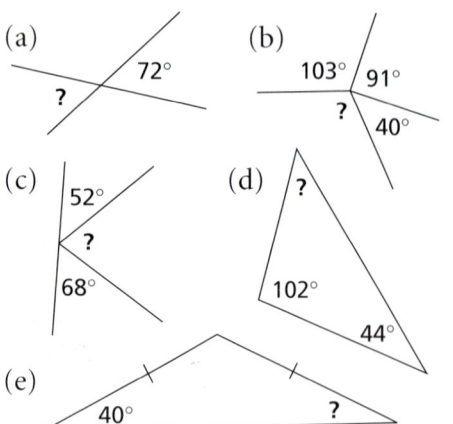

(a) 72°, ?

(b) 103°, 91°, ?, 40°

(c) 52°, ?, 68°

(d) ?, 102°, 44°

(e) 40°, ?

P Angles round a point add up to 360°.

Q Angles on a straight line add up to 180°.

R The angles of a triangle add up to 180°.

S The symmetry of an isosceles triangle gives a pair of equal sides and a pair of equal angles.

T Vertically opposite angles are equal.

When you work out angles you should be able to give a reason.

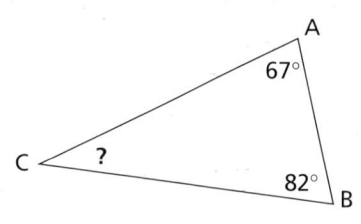

Triangle ABC: A = 67°, B = 82°, C = ?

The angle at C is 31°. These are my reasons.
I know the angles of a triangle add up to 180°.
The angles at A and B add up to 149°.
So angle C must be 180° – 149°, which is 31°.

A9 Work with a partner. Take turns to find a missing angle. Explain to your partner how you found the angle. Your partner has to be convinced by your explanation.

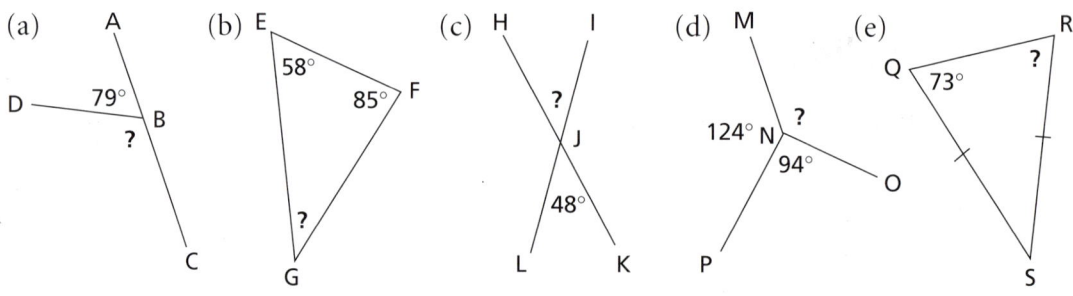

(a) A, D, B, C: 79° at A, ? at B

(b) E, F, G: 58°, 85°, ?

(c) H, I, J, K, L: ?, 48°

(d) M, N, O, P: 124°, ?, 94°

(e) Q, R, S: 73°, ?

A10 Work with a partner as in A9 to find the missing angles. With these there is more than one stage in working out the angles, so there will be more than one stage in your explanation.

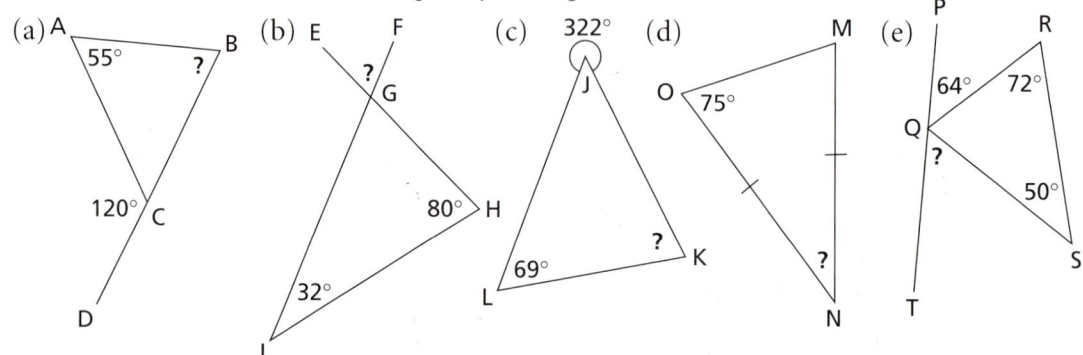

(a) A, B, C, D: 55°, ?, 120°

(b) E, F, G, H, I: ?, 80°, 32°

(c) J, K, L: 322°, 69°, ?

(d) M, N, O, K, T: 75°, ?

(e) P, Q, R, S: 64°, 72°, ?, 50°

152

A11 Work on these with a partner as in A10.

(a) Triangle with angle A = ?, angle at B (BEC) = 62°, angle DBC = 65°, with line EBC and line through B to D.

(b) Triangle EGF with angle E = 88°, two equal sides EG and EF, angle G = ?

(c) Triangle IKJ with angle I = ?, angle K = 90°, angle J = 34°, with H above I.

(d) Triangle LMP with angle L = 73°, LM and NO crossing at M, angle NMO = ?, with equal marks on LP and... (tick marks on LP and MP)

(e) Triangle TRU with angle QRT... angle QRA = 65°, angle ARS = 62° (R on line QS), angle T = 51°, angle U = ?

(f) Triangle VWX with reflex angle at W = 282°, VW = WX (tick marks), angle V = ?

A12 Shapes that are exactly the same shape and size as one another are called **congruent**.
Are any of the triangles sketched here congruent?
Give your reasons.

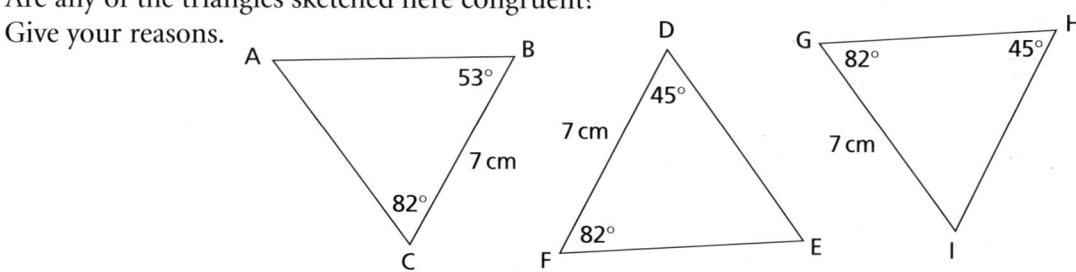

Triangle ABC: angle B = 53°, BC = 7 cm, angle C = 82°.
Triangle DEF: angle D = 45°, DF = 7 cm, angle F = 82°.
Triangle GHI: angle G = 82°, angle H = 45°, GH = 7 cm.

B Parallel lines and angles

Trace these parallel lines.

Rotate your tracing paper by part of a turn.

- How many sizes of angle can you see?
- What can you say about the positions of angles that are the same size?

You may have noticed these things about angles that were the same size in the tracing paper activity.

Some of them were **vertically opposite angles**.
You could draw an **X** over them.

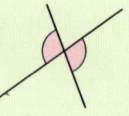

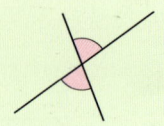

You could draw an **F-shape** over some of them.

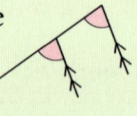

These are called **corresponding angles**.

You could draw a **Z-shape** over some of them.

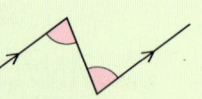

These are called **alternate angles**.

B1 Look at the angles marked with letters.
(a) Which two are a pair of vertically opposite angles?
(b) Which two are a pair of corresponding angles?
(c) Find a pair of alternate angles.
(d) Find a different pair of alternate angles.

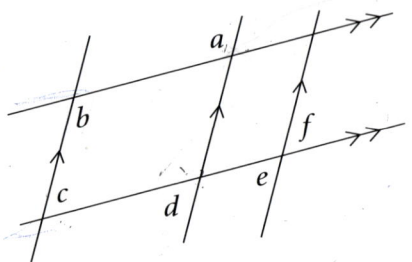

B2 Find these.
(a) A pair of alternate angles
(b) A different pair of alternate angles
(c) A pair of vertically opposite angles
(d) A pair of corresponding angles
(e) A different pair of corresponding angles

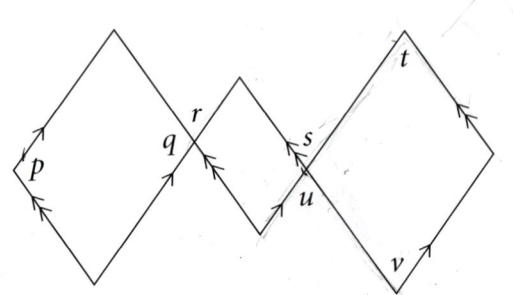

B3 For each of these, find the missing angle and say which of the three angle relationships at the top of the page gives the answer.

(a) 53°, ?
(b) ?, 42°
(c) 106°, ?
(d) ?, 45°
(e) 79°, ?
(f) 130°, ?
(g) 121°, ?
(h) 111°, ?

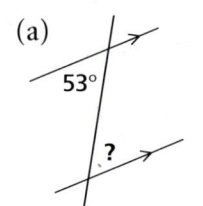

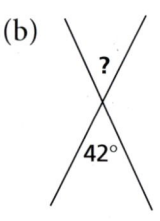

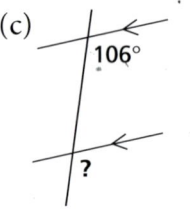

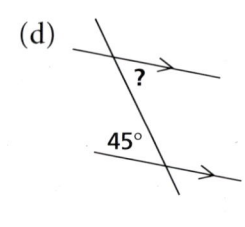

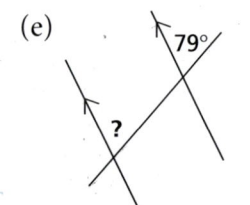

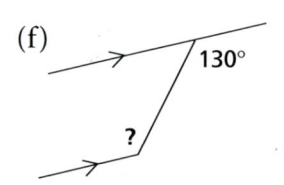

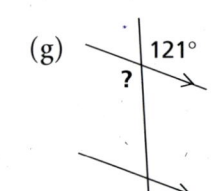

154

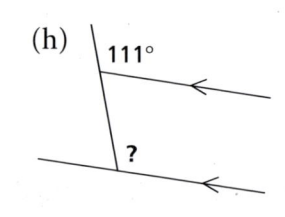

B4 For each of these, find the missing angle and say which two of these relationships you used.

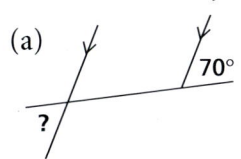

(a) 70°, ?

(b) 300°, ?

(c) 85°, 40°, ?

(d) 51°, ?

A Vertically opposite angles are equal.

B Corresponding angles are equal.

C Alternate angles are equal.

D Angles round a point add up to 360°.

E Angles on a straight line add up to 180°.

F The angles of a triangle add up to 180°.

G The symmetry of an isosceles triangle gives a pair of equal sides and a pair of equal angles.

What progress have you made?

Statement

I can find missing angles from angles I have been given.

Evidence

1 Find the missing angles.

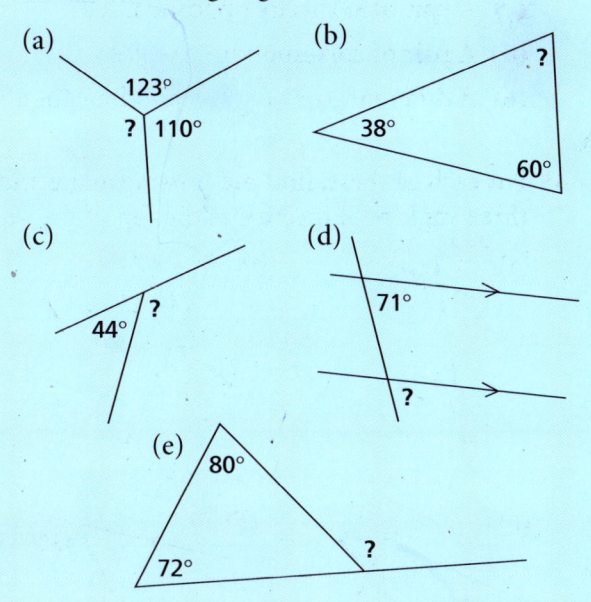

(a) 123°, ?, 110°

(b) 38°, 60°, ?

(c) 44°, ?

(d) 71°, ?

(e) 80°, 72°, ?

I can explain how I used angle relationships to get my answers.

2 Explain how you found the angles above.

155

22 Transformations

This work will help you
- transform points and shapes using translations, reflections or rotations and by using a combination of these
- describe clearly how points and shapes have been transformed

A Translations and vectors

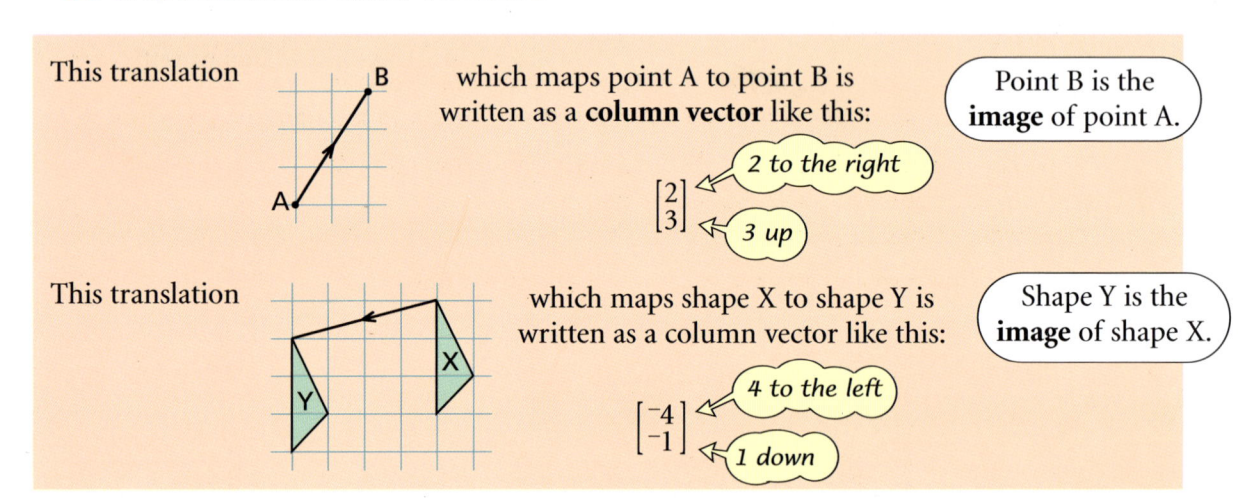

This translation which maps point A to point B is written as a **column vector** like this:

$\begin{bmatrix} 2 \\ 3 \end{bmatrix}$ — 2 to the right, 3 up

Point B is the **image** of point A.

This translation which maps shape X to shape Y is written as a column vector like this:

$\begin{bmatrix} -4 \\ -1 \end{bmatrix}$ — 4 to the left, 1 down

Shape Y is the **image** of shape X.

How will each of these translations be written as a column vector?

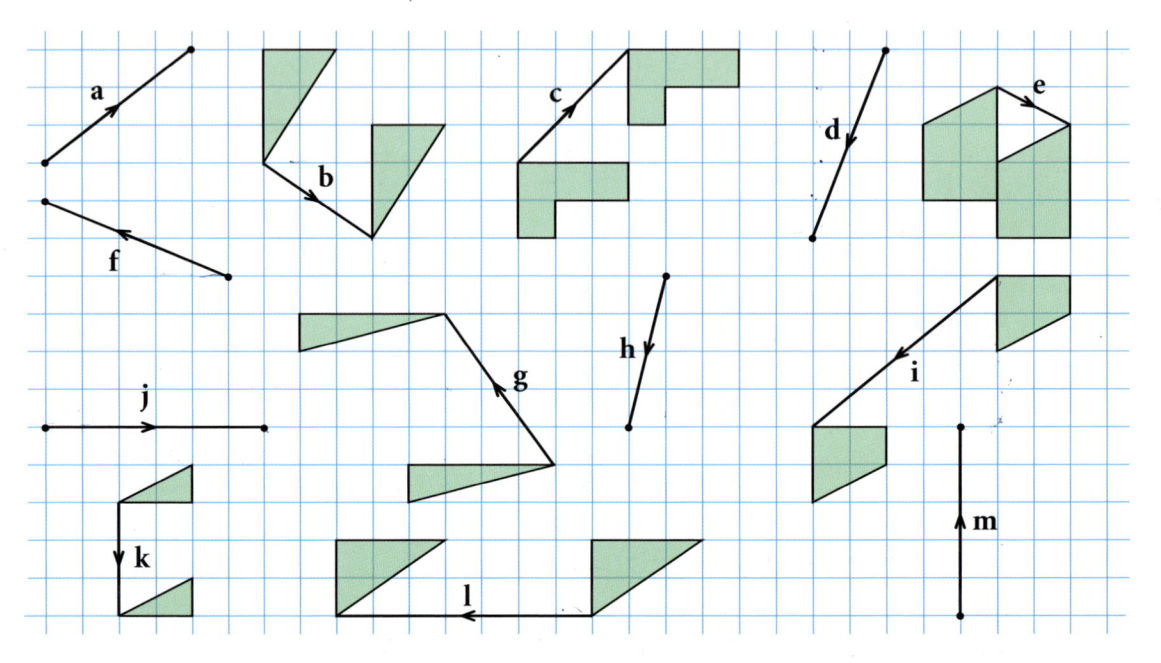

Vector snakes and ladders a game for two players

The game board is sheet 246 and the vector cards are made from sheet 247.

Each player needs a different coloured counter.

Basic game

Before you start

Put the pile of cards face down on the table.

Put both counters on START.

When it is your turn

Turn over the top card. This tells you how to move your counter.
(Your counter always goes where two grid lines cross, not in a square.)

If the move would take you off the board, you miss that go and your partner can either use that move on their go or turn over the next top card.

You slide down snakes and go up ladders in the usual way.

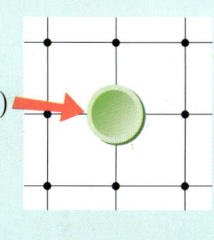

Questions A1 and A2 are about this set of L-shapes.

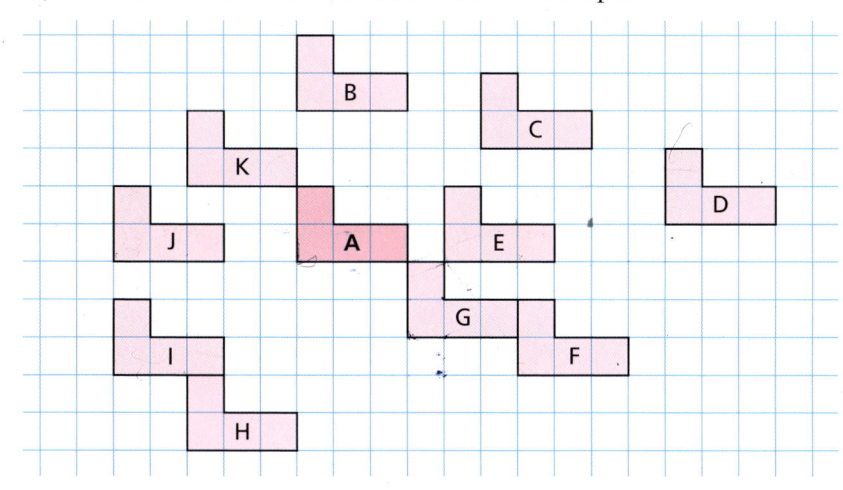

A1 Find the image of shape A after each of these translations.

(a) $\begin{bmatrix} 10 \\ 1 \end{bmatrix}$ (b) $\begin{bmatrix} 3 \\ -2 \end{bmatrix}$ (c) $\begin{bmatrix} 5 \\ 3 \end{bmatrix}$ (d) $\begin{bmatrix} 0 \\ 4 \end{bmatrix}$ (e) $\begin{bmatrix} -3 \\ 2 \end{bmatrix}$

(f) $\begin{bmatrix} 6 \\ -3 \end{bmatrix}$ (g) $\begin{bmatrix} -3 \\ -5 \end{bmatrix}$ (h) $\begin{bmatrix} 4 \\ 0 \end{bmatrix}$ (i) $\begin{bmatrix} -5 \\ 0 \end{bmatrix}$ (j) $\begin{bmatrix} -5 \\ -3 \end{bmatrix}$

A2 Use column vectors to describe the translations that map

(a) G to D (b) E to F (c) E to I

(d) I to J (e) K to H (f) G to K

A3 (a) Draw a set of axes numbered from 0 to 10 along each axis.
Plot the points (1, 1), (3, 1) and (1, 6).
Join them up and label the triangle A.

(b) Draw the image of triangle A after a translation of $\begin{bmatrix} 6 \\ 1 \end{bmatrix}$.
Label this image B.

(c) Draw the image of triangle A after a translation of $\begin{bmatrix} 2 \\ 4 \end{bmatrix}$.
Label this image C.

(d) What translation will map triangle C to B?

(e) What translation will map triangle B to C?

B Reflection

B1 (a) Make a copy of triangle ABC.

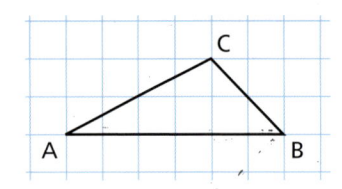

Reflect the triangle in the line AB and draw the image.
What is the name of the shape formed by triangle ABC and its image?

(b) Make another copy of triangle ABC.
Reflect the triangle in the line BC and draw the image.
What is the name of the shape formed by the triangle and its image?

B2 (a) Shape D can be reflected on to two of these shapes.
What are the two shapes?

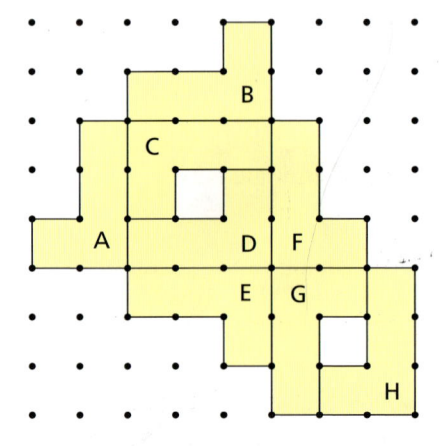

(b) Find as many pairs of shapes as you can that are mirror images of each other.

B3 Answer this on sheet 265.

B4 Draw a grid like this with each axis numbered from −8 to 8.

(a) (i) Plot the points (6, 4), (7, 3), (8, 4) and (8, 6). Join them up and label the shape A.

(ii) Reflect shape A in the y-axis. Label the image A'.

(iii) Reflect shape A in the line $y = x$. Label the image A".

(b) (i) Plot the points (3, 1), (3, 2) and (7, 1). Join them up and label the shape B.

(ii) Reflect shape B in the x-axis. Label the image B'.

(iii) Reflect shape B in the line $y = x$. Label the image B".

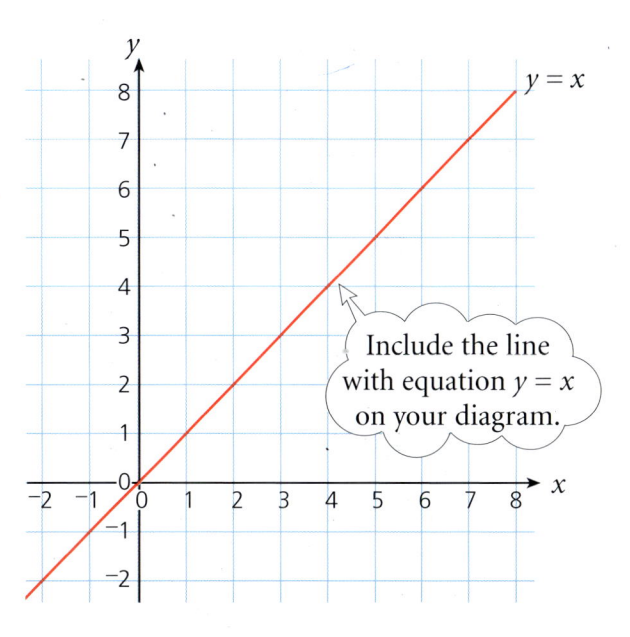

Include the line with equation $y = x$ on your diagram.

C Rotation

To describe a rotation fully, you need
- the amount of turn (the angle)
- the direction (clockwise or anticlockwise)
- the centre of rotation

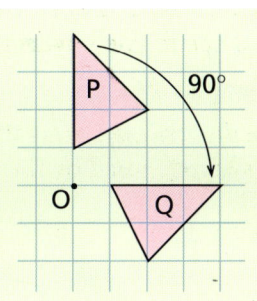

A rotation that maps shape P to Q is a rotation of 90° clockwise about the point O.

C1 (a) Make a copy of triangle ABC.

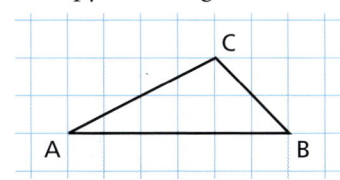

Draw the image of the triangle after a 90° turn anticlockwise about B.

(b) Make another copy of triangle ABC.
Mark the point half-way along the line AB and label it X.

Draw the image of the triangle after a 180° turn about X.
What is the name of the shape formed by the triangle and its image?

159

C2 (a) Draw a pair of axes, both numbered from ⁻5 to 5.
Plot the points (2, 1), (5, 1) and (5, 3).
Join them up and label the triangle P.

(b) Draw the image of P after a rotation of 180° about the point (0, 0).
Label this image Q.

(c) Draw the image of P after a rotation of 90° anticlockwise about the point (0, 0).
Label this image R.

(d) What rotation will map triangle Q to R?

This repeating pattern is from the Alhambra Palace at Granada in Spain.
Some points and shapes have been labelled.

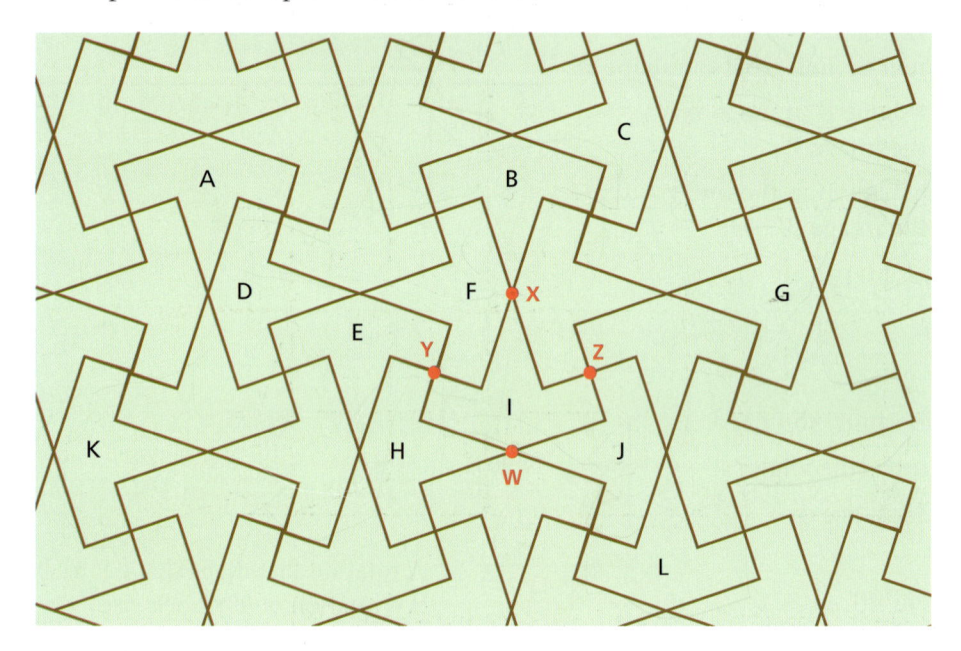

C3 Which points are the centres of rotations that map (a) B to I (b) I to F

C4 (a) What is the image of shape J after a rotation of 180° about the point W?
(b) Find the image of shape E after an anticlockwise rotation of 90° about the point Y.

C5 Describe **fully** the rotation that maps shape I to shape J.

C6 Which points are the centres of rotations that map (a) D to G (b) L to B

C7 (a) What is the image of shape D after a rotation of 180° about the point Y?
(b) What is the image of shape G after a clockwise rotation of 90° about the point Z?

C8 Describe **fully** a rotation that maps
(a) B to L (b) E to L (c) D to B (d) C to A

D Translation, reflection and rotation

The repeating pattern below is based on a frieze on a marble floor at the Mausoleum of Mohammed V at Rabat in Morocco.

Some shapes have been labelled and some points and lines are shown to help describe any transformations.

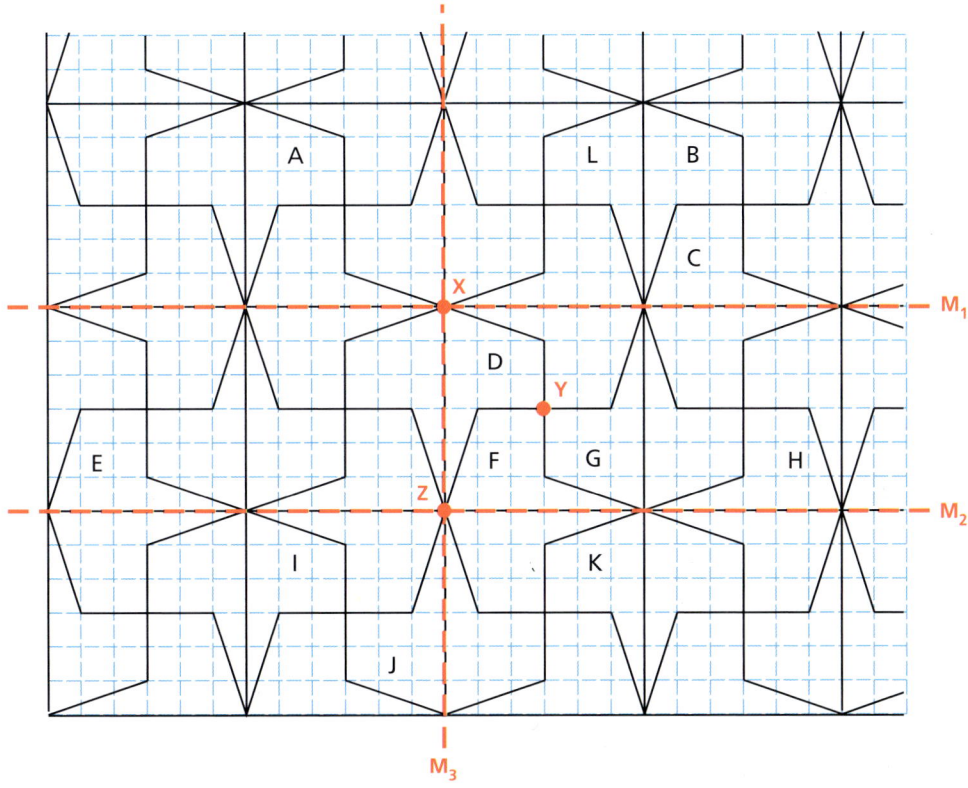

D1 Which shape is the image of (a) shape E after the translation $\begin{bmatrix} 18 \\ 6 \end{bmatrix}$

(b) shape G after reflection in the line M_2

(c) shape J after a rotation of 180° about point Y

(d) shape F after a clockwise rotation of 90° about Y

D2 (a) Describe a reflection that maps shape I to shape K.

(b) Describe a translation that maps shape A to shape B.

(c) Describe a rotation that maps shape G to shape D.

(d) Describe a rotation that maps shape G to shape F.

(e) Describe a rotation that maps shape A to shape G.

D3 Describe fully a transformation that maps

(a) I to B (b) A to L (c) D to F (d) I to G

E Combining transformations

E1 (a) Make a copy of triangle A.

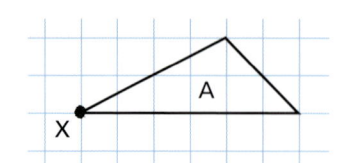

   Draw the image of the triangle after a rotation of 90° anticlockwise about X. Label this image A_1.

(b) Now draw the image of A_1 after another rotation of 90° anticlockwise about X. Label this image A_2.

(c) Finally draw the image of A_2 after another rotation of 90° anticlockwise about X. Label this image A_3.

(d) Describe a single transformation that will map triangle A to triangle A_2.

(e) Describe a single transformation that will map triangle A to triangle A_3.

E2 Mark a point P on some squared paper.

(a) From P, draw a journey consisting of these vectors.

$\begin{bmatrix} 3 \\ 0 \end{bmatrix}$, then $\begin{bmatrix} 2 \\ 1 \end{bmatrix}$, then $\begin{bmatrix} 1 \\ 3 \end{bmatrix}$

Mark the finishing point Q.

(b) Start again at the point P you marked.
This time draw the vectors in this order.

$\begin{bmatrix} 1 \\ 3 \end{bmatrix}$, $\begin{bmatrix} 2 \\ 1 \end{bmatrix}$, $\begin{bmatrix} 3 \\ 0 \end{bmatrix}$

Where do you finish?

(c) What happens when you start from P and use the vectors in this order?

$\begin{bmatrix} 2 \\ 1 \end{bmatrix}$, $\begin{bmatrix} 3 \\ 0 \end{bmatrix}$, $\begin{bmatrix} 1 \\ 3 \end{bmatrix}$

E3 Mark points M and N carefully on some squared paper.
You have to draw a journey from M to N.
You must use $\begin{bmatrix} 4 \\ 3 \end{bmatrix}$, $\begin{bmatrix} 1 \\ -2 \end{bmatrix}$ and one other vector.

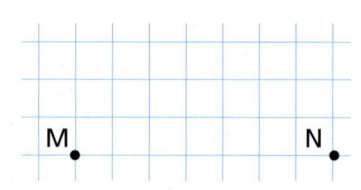

What is this other vector?

E4 You want to translate shape R on to shape S.
You can use three out of these four vectors.

$$\begin{bmatrix} 4 \\ 2 \end{bmatrix}, \begin{bmatrix} 2 \\ 1 \end{bmatrix}, \begin{bmatrix} 2 \\ 3 \end{bmatrix}, \begin{bmatrix} 3 \\ 1 \end{bmatrix}$$

(a) Which three vectors do you need?
(b) Does it matter which order you use them?

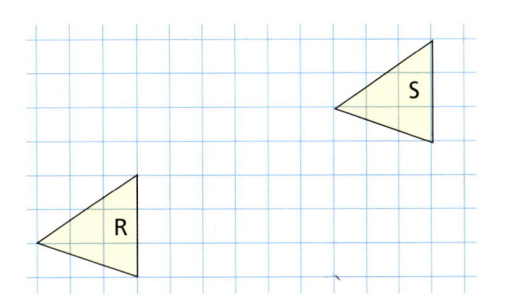

E5 (a) You want to translate shape C to shape D.
Use two out of these four vectors.

$$\begin{bmatrix} 1 \\ -2 \end{bmatrix}, \begin{bmatrix} 3 \\ 2 \end{bmatrix}, \begin{bmatrix} 4 \\ 1 \end{bmatrix}, \begin{bmatrix} 2 \\ -1 \end{bmatrix}$$

(b) You can get from C to D using three out of these five vectors. Which ones do you need?

$$\begin{bmatrix} -1 \\ -2 \end{bmatrix}, \begin{bmatrix} 3 \\ 1 \end{bmatrix}, \begin{bmatrix} -1 \\ 1 \end{bmatrix}, \begin{bmatrix} -2 \\ -2 \end{bmatrix}, \begin{bmatrix} 7 \\ 0 \end{bmatrix}$$

(c) What single vector would take you straight from C to D?

E6 What single translation would have the same effect as translating by $\begin{bmatrix} 5 \\ 2 \end{bmatrix}$ then $\begin{bmatrix} 1 \\ 3 \end{bmatrix}$?

E7 Translating by $\begin{bmatrix} 4 \\ a \end{bmatrix}$ then $\begin{bmatrix} b \\ -2 \end{bmatrix}$ has the same effect as translating by $\begin{bmatrix} 7 \\ 3 \end{bmatrix}$.

Find the values of a and b.

E8 Copy the diagram on to squared paper.

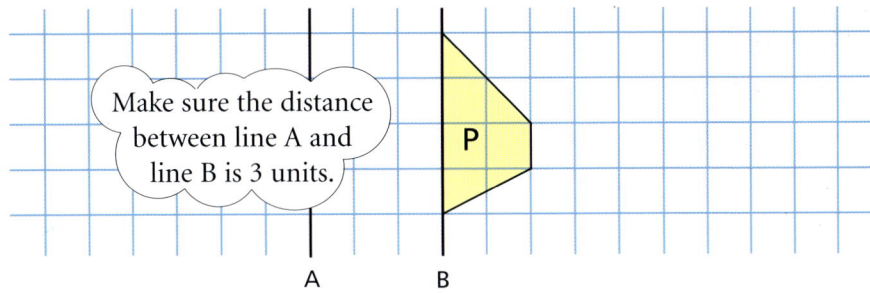

Make sure the distance between line A and line B is 3 units.

(a) (i) Reflect shape P in line B and label the image P_1.
  (ii) Reflect P_1 in line A and label the final image P_2.
(b) Make another copy of triangle P and lines A and B.
 Now repeat (a) but begin by reflecting shape P in line A and then line B.
(c) Does it matter in which order you do these reflections?

E9 Copy the shape and lines A and B on to the middle of a piece of squared paper.

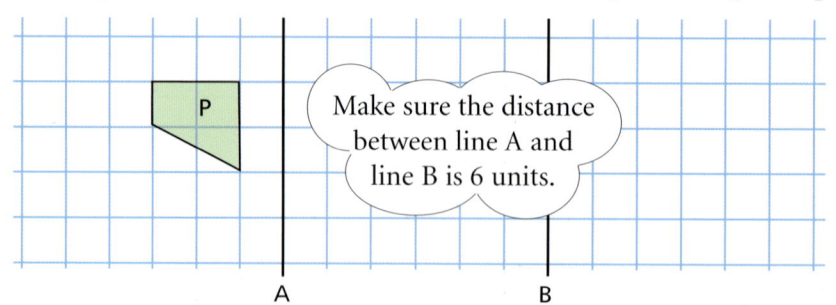

Make sure the distance between line A and line B is 6 units.

(a) (i) Reflect shape P in line A and label the image P_1.

(ii) Reflect P_1 in line B and label the final image P_2.

(iii) What single transformation could you use to map P on to P_2?

(b) Now draw your own shape on the diagram.
Reflect it in line A and then line B.

Try this with different shapes.
What do you notice?

E10 (a) Draw a pair of axes numbered from ⁻6 to 6 along each axis.
Plot the points (3, 1), (2, 4), (4, 5) and (5, 4).
Join them up and label the shape A.

(b) (i) Draw the image of shape A after a reflection in the x-axis.
Label the image A'.

(ii) Draw the image of shape A' after a reflection in the y-axis.
Label the image A".

(c) Now repeat (b) but begin by reflecting shape A in the y-axis and then the x-axis.

(d) Does it matter in which order you do these reflections?

(e) What single transformation could you use to map shape A on to A"?

E11 Draw a grid like this with the x-axis numbered from ⁻5 to 5 and the y-axis from 0 to 5.

(a) Plot the points (1, 2), (1, 5) and (3, 5).
Join them up and label the shape X.

(b) Transform shape X by reflecting in the line $y = x$ followed by reflecting in the y-axis.
Label the final image Y.

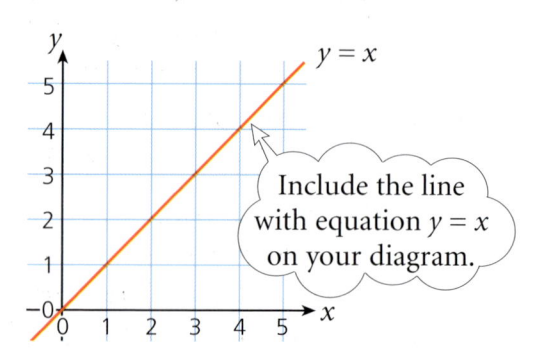

Include the line with equation $y = x$ on your diagram.

(d) What single transformation could you use to map shape X on to Y?

Questions E12 to E15 are about the set of shapes below.

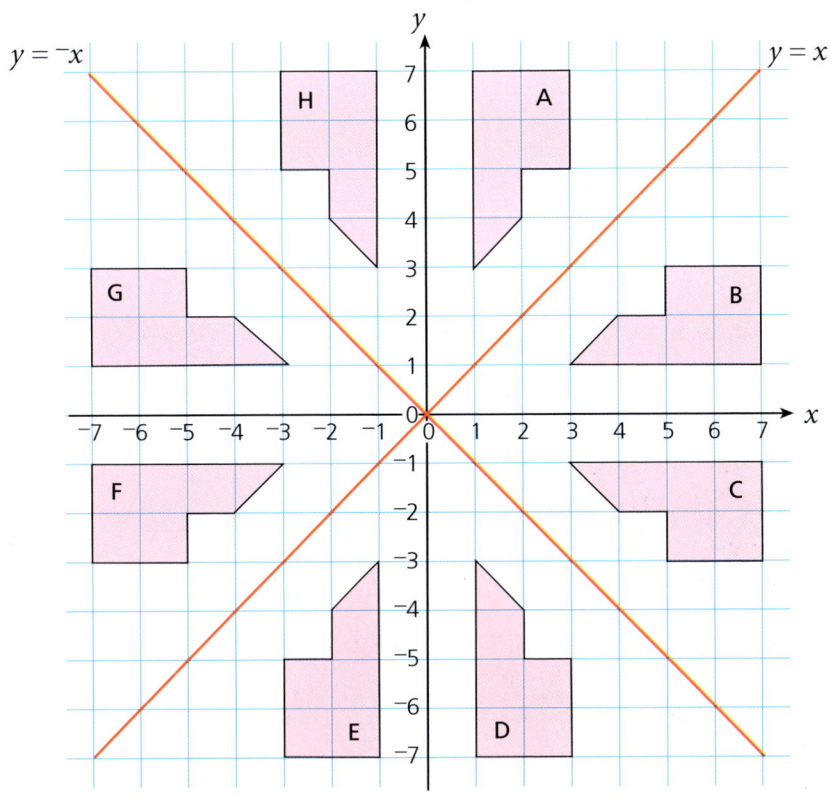

E12 (a) (i) Reflect shape A in the y-axis and then rotate the image 180° about (0, 0). What shape is the final image of A?

(ii) Repeat (i) for a different shape on the grid.

(b) What single transformation could replace a reflection in the y-axis followed by a rotation of 180° about (0, 0)?

E13 (a) (i) What is the image of shape C after a reflection in the line $y = ^-x$ followed by a clockwise rotation of 90° about (0, 0)?

(ii) Repeat (i) for a different shape on the grid.

(b) What single transformation could replace a reflection in the line $y = ^-x$ followed by a clockwise rotation of 90° about (0, 0)?

E14 (a) (i) What is the image of shape C after a clockwise rotation of 90° about (0, 0) followed by a reflection in the x-axis?

(ii) Repeat (i) for a different shape on the grid.

(b) What single transformation could replace a clockwise rotation of 90° about (0, 0) followed by a reflection in the x-axis?

E15 Write down a **pair** of transformations that will map

(a) D to F   (b) C to G   (c) E to H   (d) H to C

What progress have you made?

Statement

I can carry out and describe translations, reflections and rotations, and combinations of these.

Evidence

Some identical shapes are shown on the grid. Some points and lines have been labelled.

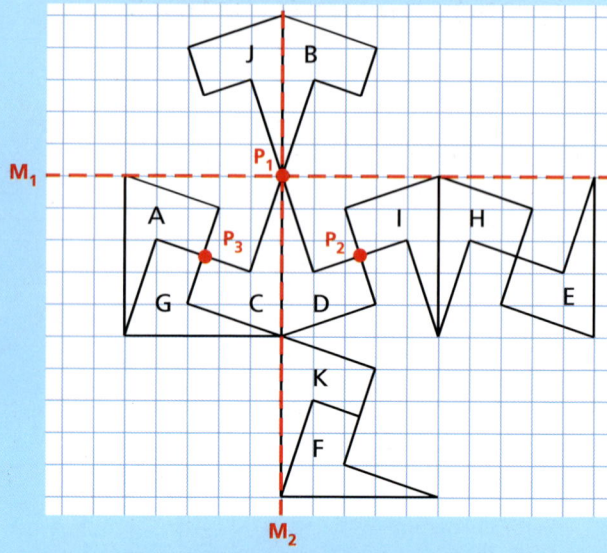

1 To translate shape K onto shape H you can use two out of these four vectors.

$$\begin{bmatrix} 6 \\ 3 \end{bmatrix}, \begin{bmatrix} 4 \\ 3 \end{bmatrix}, \begin{bmatrix} 1 \\ -1 \end{bmatrix}, \begin{bmatrix} -1 \\ 2 \end{bmatrix}$$

Which two vectors can you use?

2 (a) What is the image of shape B after a rotation of 180° about the point P_1?

(b) What is the image of shape F after a rotation of 90° anticlockwise about P_2?

3 Describe fully transformations that map

(a) A to I   (b) A to E   (c) G to A

4 (a) What is the image of C after reflection in M_1 followed by a rotation of 180° about P_1?

(b) What is the image of D after a rotation of 90° anticlockwise about P_2 followed by a reflection in M_2?

5 Write down a pair of transformations that will map shape C onto shape I.

23 Trial and improvement

This will help you solve problems using trial and improvement methods.

A Introducing the method

Problem 1

I think of a number. I add 12.
I multiply the result by 13.
The answer is 429.

- What number did I start with?

Problem 2

I think of a number. I add 7.
I multiply the result by my starting number.
The answer is 408.

- What number did I start with?

Problem 3

I think of two consecutive numbers.
I add them together.
The answer is 201.

- What numbers did I start with?

Problem 4

I think of two consecutive numbers.
I multiply them together.
The answer is 1806.

- What numbers did I start with?

A1 I think of a number. I subtract 9.
I multiply the result by 14.
The answer is 266.

What number did I start with?

A2 I think of a number. I add 8.
I multiply the result by my starting number.
The answer is 1833.

What number did I start with?

A3 I multiply together two consecutive numbers and my answer is 2352.
What numbers did I start with?

A4 Find three consecutive numbers that multiply together to give 74 046.

A5 This rectangle is 3 times as long as it is wide.
  (a) If its area is 192 cm^2, how long and how wide is it?
  (b) If its area is 1323 cm^2, how long and how wide is it?

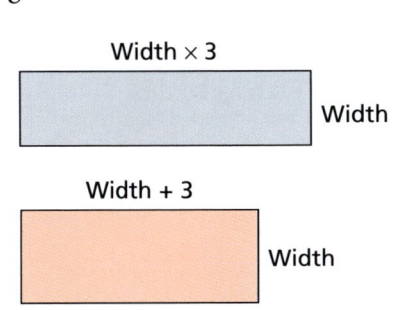

A6 The length of this rectangle is 3 cm more than its width.
If its area is 418 cm^2, how long and how wide is it?

A7 $n(n + 2) = 14883$ and n is a positive whole number. Find n.

A8 Find n where $n^3 = 4096$.

A9 Find n where $n^2 + n = 812$ and n is a positive number.

B Being systematic

Problem 1

Two positive whole numbers differ by 5 and multiply to make 5106. Find the numbers.

First number	Second number	Multiply together	Result Too small	Too big
30	35	1050	✔	
50	55	2750	✔	
73	78	5694		✔
71	76	5396		✔
69	74	5106	Exactly right!	

Problem 2

Solve the equation $n^2 + n = 75.44$ where n is a positive number.

n	$n^2 + n$	Result Too small	Too big
7	56	✔	
9	90		✔
8	72	✔	
8.5	80.75		✔

B1 Two positive whole numbers differ by 3 and multiply to give 2548. Find the numbers.

You could show your trials in a table like this.

First number	Second number	Multiply together	Result Too small	Too big
33	36	1188	✔	

B2 Solve the equation $n^2 - n = 4692$ where n is a positive number.

B3 Solve the equation $n(n + 1) = 38.19$ where n is a positive number.
You could show your trials in a table like this.

| | | Result | |
n	n(n + 1)	Too small	Too big
6	42		✓

B4 Two positive numbers differ by 7 and multiply to give 81.84.
Find the numbers.

B5 A rectangle has an area of $104.5\,\text{cm}^2$.
Its length is 1.5 cm more than its width.

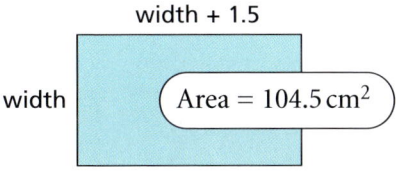

width + 1.5
width
Area = $104.5\,\text{cm}^2$

Copy and complete the table below to find the width and length of this rectangle.

| | Length (cm) | | Result | |
Width (cm)	(width + 1.5)	Area (cm²)	Too small	Too big
7	8.5	59.5		

B6 A rectangle has an area of $38.88\,\text{cm}^2$.
Its length is 3 times its width.

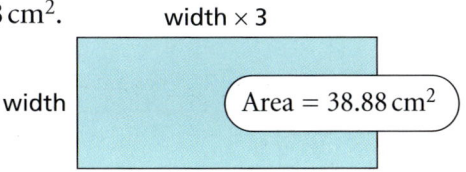

width × 3
width
Area = $38.88\,\text{cm}^2$

Find the width and length of this rectangle.

A contest between two people

12 768 tiles, each 1 cm square can be arranged in a rectangle whose length is 2 cm more than its width. Find the length and width of the rectangle.

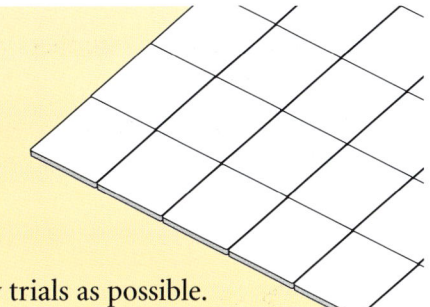

Each of you finds the solution to this problem using as few trials as possible.
Who did it in fewer trials, you or your partner?

C Not exactly

Problem

A rectangle has an area of 24 cm². Its length is 3 cm more than its width. What is the width of the rectangle?

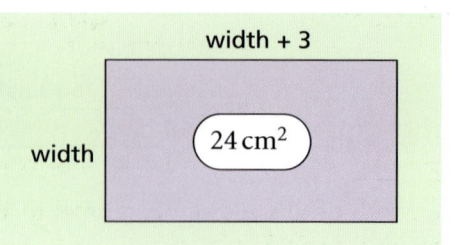

- A width of 3 cm gives an area of 18 cm².

  Too small

- A width of 4 cm gives an area of 28 cm².

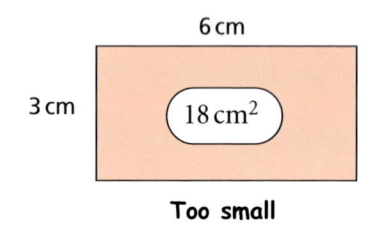

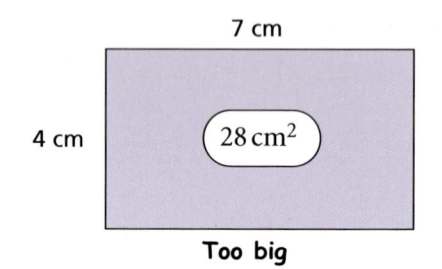

  Too big

- The width must be between 3 cm and 4 cm.

- We can show our trials in a table …   … or on a number line.

Width (cm)	Length (cm)	Area (cm²)	Too small/ Too big
3	6	18	Too small
4	7	28	Too big
3.5	6.5	22.75	Too small
3.6	6.6	23.76	Too small
3.7	6.7	24.79	Too big
3.62	6.62	23.9644	Too small
3.63	6.63	24.0669	Too big

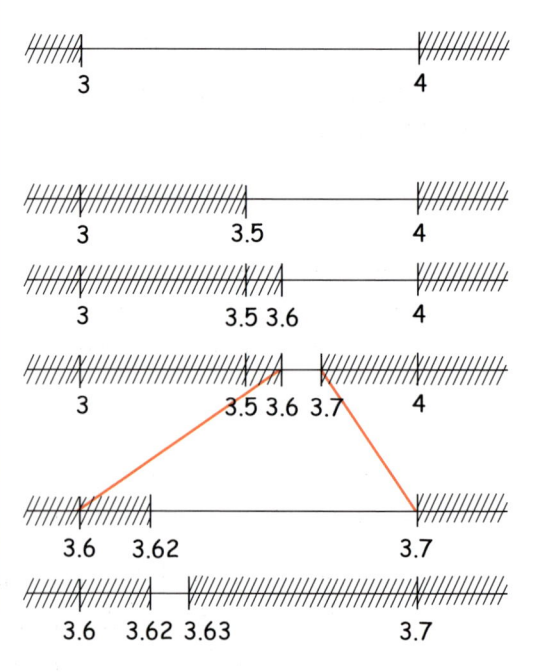

- The decimal value for the width begins 3.62…… so the width is 3.6 cm to 1 d.p.

C1 A rectangle's length is 4 cm more than its width. Its area is 100 cm².

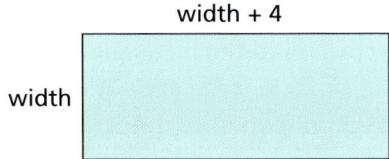

Find the width of this rectangle, correct to 1 d.p. Start by trying a width of 9 cm.

C2 A rectangle's length is 5 times its width. Its area is 100 cm².

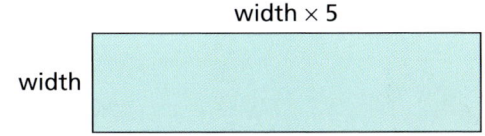

Find the width of this rectangle, correct to 1 d.p.

C3 Find a solution (correct to 1 d.p.) of the equation

$$x^2 + x = 10$$

You could use a table like this to record your results.

x	$x^2 + x$	Result Too small	Too big
2			
3			

C4 Find the solution of the equation $n^3 - n = 5$, correct to 1 d.p. Start by trying $n = 2$.

C5 Find the solution of the equation $x^3 + x = 40$, correct to 1 d.p.

C6 T - Can you make me a cuboid please? The width must be half the length, and the height must be half the width. Volume to be 150 cm³. Ta - J

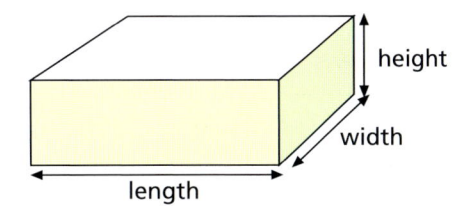

Work out the dimensions of Jenny's cuboid correct to 1 d.p. You could use a table like this to record your results.

Length (cm)	Width (cm)	Height (cm)	Volume (cm³)	Result Too big	Too small
8	4	2			

D Using a spreadsheet

A rectangle has its length four times its width.
The area of the rectangle is 280 cm². Find the rectangle's dimensions.

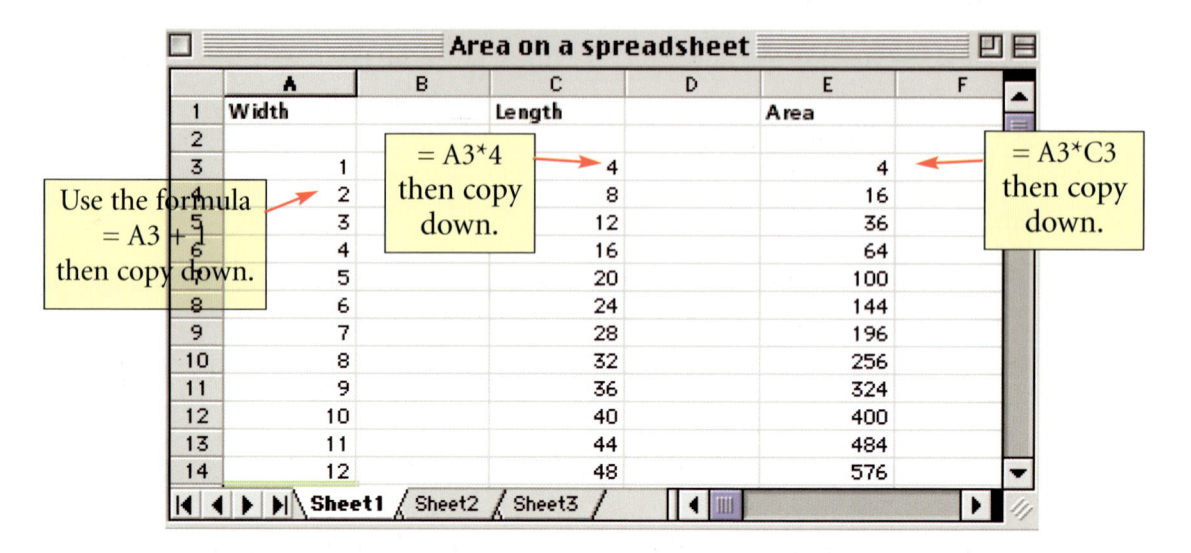

Area on a spreadsheet

	A	B	C	D	E	F
1	Width		Length		Area	
2						
3	1		4		4	
4	2		8		16	
5	3		12		36	
6	4		16		64	
7	5		20		100	
8	6		24		144	
9	7		28		196	
10	8		32		256	
11	9		36		324	
12	10		40		400	
13	11		44		484	
14	12		48		576	

Use the formula $= A3 + 1$ then copy down.

$= A3*4$ then copy down.

$= A3*C3$ then copy down.

You can see that a width of 8 gives area 256 cm² – too small
and a width of 9 gives area 324 cm² – too big.

So the width we want is between 8 cm and 9 cm.

Area on a spreadsheet

	A	B	C	D	E	F
	Width		Length		Area	
3	8		32		256	
4	8.1		32.4		262.44	
5	8.2		32.8		268.96	
6	8.3		33.2		275.56	
7	8.4		33.6		282.24	
8	8.5		34		289	
9	8.6		34.4		295.84	
10	8.7		34.8		302.76	
11	8.8		35.2		309.76	
12	8.9		35.6		316.84	
13	9		36		324	
14	9.1		36.4		331.24	

Change the value in A3 to 8.

Change the formula to $= A3 + 0.1$ and copy down.

Now you can see that the width you want is between 8.3 and 8.4 cm.

- Change the value in cell A3 to 8.3.
  Change the formula in cell A4 to go up in steps of 0.01.
- Between what two numbers must the width lie?
  Carry on until you can be sure of the width to 2 d.p.

D1 The length of this rectangle is 11 cm more than the width. Its area is 200 cm².

Use a spreadsheet to find the width of this rectangle, accurate to 2 d.p.

D2 To find the length of this rectangle you take the width, multiply it by 2 and then add 1. It has an area of 100 cm².

Use a spreadsheet to find the width of this rectangle, accurate to 2 d.p.

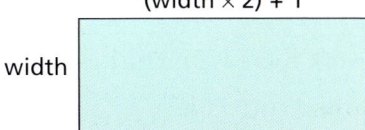

D3 Use trial and improvement on a spreadsheet to find the cube root of 150, correct to 2 d.p.

You will only need to use two columns such as:

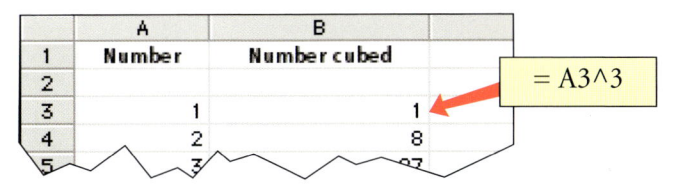

	A	B
1	Number	Number cubed
2		
3	1	1
4	2	8
5	3	27

= A3^3

D4 Use a spreadsheet to find a positive solution to $n^2 + n = 300$, correct to 2 d.p.

You will only need to use two columns such as:

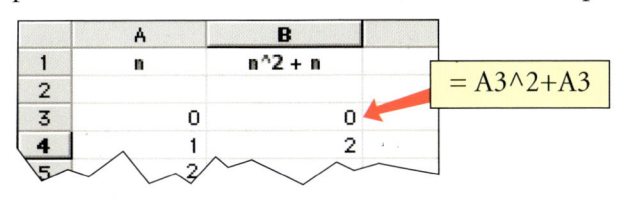

	A	B
1	n	n^2 + n
2		
3	0	0
4	1	2
5	2	

= A3^2+A3

D5 Use a spreadsheet to find a solution of $x^2 - x = 150$, correct to 2 d.p.

D6 Use a spreadsheet to find the cube root of 500, correct to **3 d.p.**

D7 Use a spreadsheet to find the solution of $x^3 + 2x = 10$, correct to 3 d.p.

What progress have you made?

Statement

I can use trial and improvement to solve problems.

Evidence

1 Find three consecutive whole numbers that multiply together to give 35 904.

2 A rectangle has a length that is 1 cm more than its width. Its area is 25 cm². Find its width, correct to 1 decimal place.

3 Find a solution of the equation $x^2 - x = 40$, correct to 1 d.p. Start by trying $x = 6$.

24 Spot the errors

> This work will help you
> - recognise when you are making these mistakes in your own work
> - remember to check your own working and to make sure that answers are sensible.

In each piece of work there is one mistake.

Spot each mistake and correct it.

Try to say how the mistake was made.

1 What fraction of the rectangle is shaded?

The rectangle is split into 3 pieces.

One piece is shaded so $\frac{1}{3}$ is shaded.

2 Claire's class went on a trip in minibuses.
There are 25 children in her class.
Each minibus holds 10 children.
How many minibuses were needed?

$25 \div 10 = 2.5$

so 2.5 minibuses were needed

3 Write in figures,
three thousand and forty eight.

300048

4 Write 4 metres 7 centimetres in metres.

4.7 metres

5 What is the area of this rectangle?

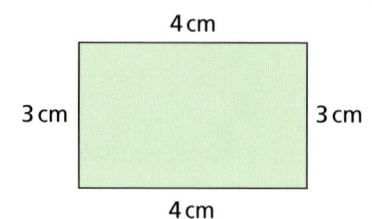

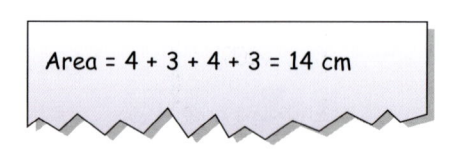

4 cm

3 cm   3 cm

4 cm

Area = 4 + 3 + 4 + 3 = 14 cm

6 Andrew bought 10 packets of sweets for £5.
How much did each packet cost?

$10 \div 5 = 2$

so each packet cost £2

7 Calculate 12.6 + 3

$$\begin{array}{r} 12.6 \\ + \ 3 \\ \hline 12.9 \end{array}$$

8 Change 2000 metres to kilometres.

2000 ÷ 100 = 20

9 A bag of sugar weighs 1 kg and 20 grams of this sugar is lost through a hole in the bottom. How much sugar is left in the bag?

100 − 20 = 80

so 80 grams are left in the bag

10 Which of these numbers are prime?
3, 4, 5, 6, 7, 8, 9, 10, 11, 12, 13, 14

3, 5, 7, 9, 11, 13

11 Find the missing number in ■ − 9 = 3.

9 − 6 = 3 so the missing number is 6

12 Calculate: (a) 10% of £60
(b) 5% of £40

(a) 60 ÷ 10 = 6 so answer is £6

(b) 40 ÷ 5 = 8 so answer is £8

13 Bars of chocolate cost £0.65 each. What is the cost of 8 chocolate bars?

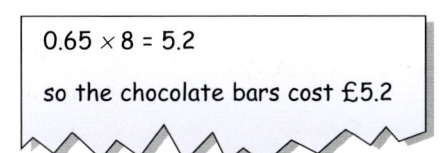

0.65 × 8 = 5.2

so the chocolate bars cost £5.2

14 Calculate 37 × 29.

$$\begin{array}{r} 37 \\ \times \ 29 \\ \hline 333 \\ 74 \\ \hline 407 \end{array}$$

15 Calculate $\frac{20 + 8}{4}$.

20 + 8 ÷ 4 = 22

16 Calculate $10 - 2 \times 3$.

10 − 2 × 3 = 24

17 The time is now 3:00.
What was the time 50 minutes ago?

3:00 − 50 = 2:50

18 Put these numbers in order, smallest first.
0.03, 0.2, 0.5, 0.004, 0.0065

0.2, 0.03, 0.004, 0.5, 0.0065

19 From the list of numbers, 0.07, 0.05, 0.129, 0.19, 0.23, 0.085, which is the largest number?

0.129

20 Draw a rectangle.
Show clearly any lines of symmetry.

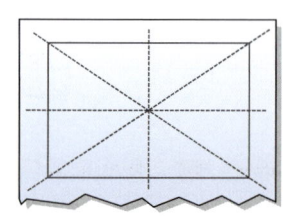

21 This bar chart shows the favourite colours for class 9Y.
What is the mode for this chart?

The tallest bar shows 9 people like blue best so the mode is 9.

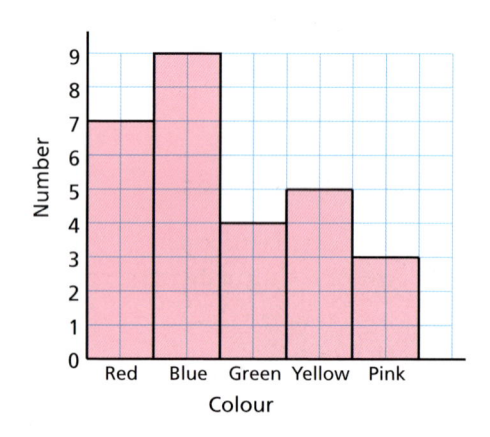

22 Can you decide on the probability that it will be wet tomorrow?

Yes. It will be wet or dry tomorrow so the probability it will be wet is $\frac{1}{2}$.

23 What is the area of this shape?

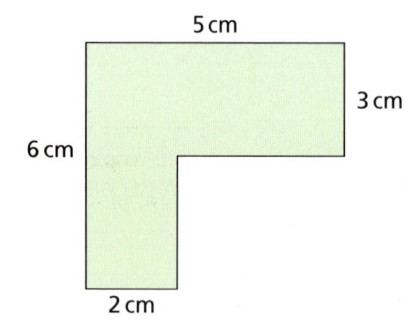

Area = 5 × 3 × 2 × 6 = 180 cm²

24 Measure this angle in degrees.

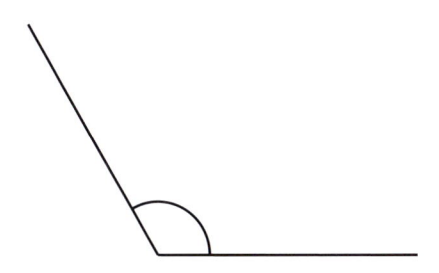

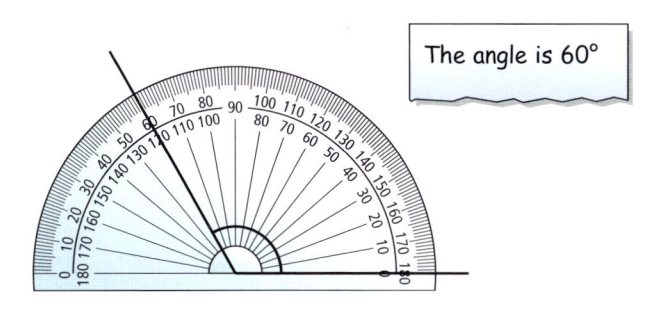

The angle is 60°

25 Draw a parallelogram.
Show clearly any lines of symmetry.

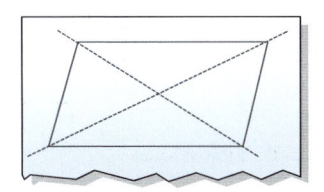

26 The pointer in this spinner is spun round.
What is the probability of
the arrow pointing to red?

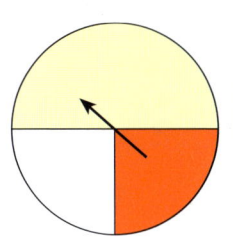

Probability of red is $\frac{1}{3}$.

27 Find the volume of this cuboid.

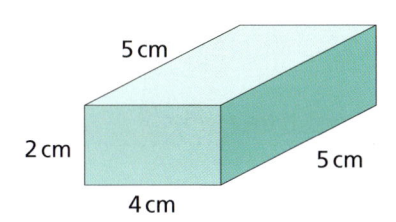

The volume is 2 × 4 × 5 × 5 = 200 cm³

28 What is the area of this triangle?

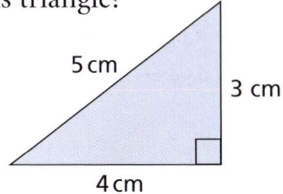

The area is $\frac{5 \times 4 \times 3}{2}$ = 30 cm²

29 Measure this angle in degrees.

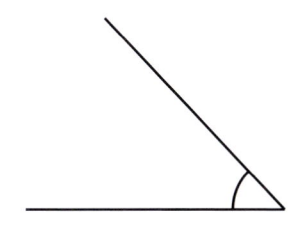

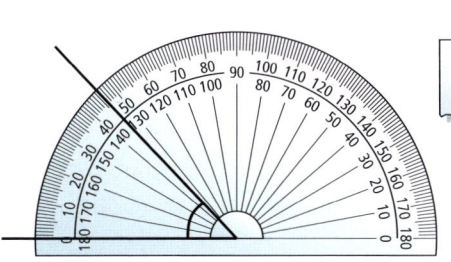

The angle is 54°

177

25 Stretchers

These are challenges where you have to use your own methods.

A Darts

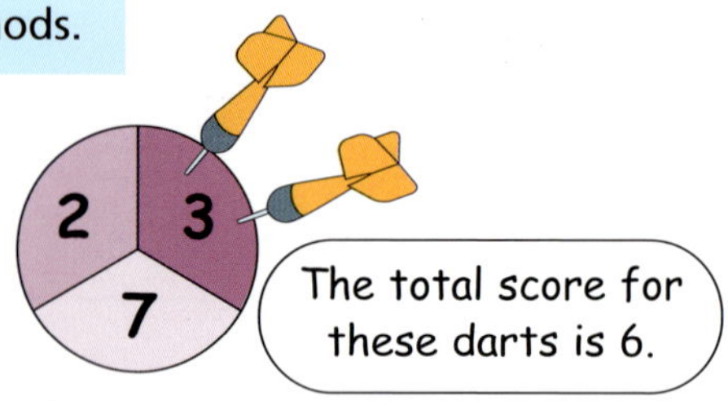

A1 Two darts are thrown at this dartboard.

The total score for these darts is 6.

(a) What are the total scores for these dartboards?

(i)   (ii)

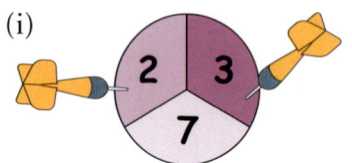

(b) Write down all the possible totals when two darts hit this dartboard.

A2 What are all the possible totals with two darts for each of these dartboards?

(a) 4, 1, 2 (b) 2, 3, 4 (c) 1, 3, 2 (d) 6, 1, 1

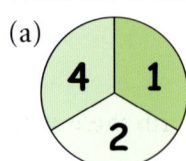

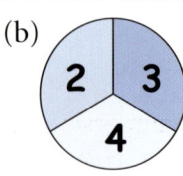

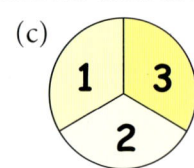

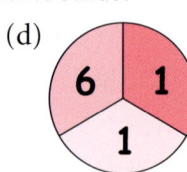

A3 (a) On a dartboard with three sectors the totals you can get with two darts are

    10   13   16   19   22

What numbers are on this board?

(b) On a similar dartboard the totals you can get with two darts are

    4   5   6   7   8   10

What numbers are on this board?

(c) On another similar dartboard the totals you can get with two darts are

    4   11   18

What numbers are on this board?

A4 Two darts are thrown at this dartboard. Write down all the possible totals with two darts for this dartboard.

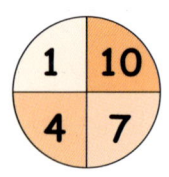

(dartboard with sectors 1, 10, 4, 7)

A5 (a) Design a dartboard with four sectors that gives three different possible totals with two darts.

(b) Design a dartboard with four sectors that gives ten different possible totals with two darts.

B Minimal measuring

Here is a set of four strips.

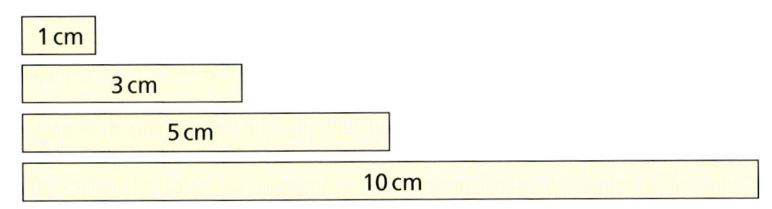

[1 cm]
[3 cm]
[5 cm]
[10 cm]

B1 You can make different lengths by putting some of these strips end to end like this.

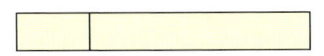

(a) Which strips could you use to make a length of 13 cm?
(You cannot use a strip more than once.)

(b) List all the different lengths you can make by putting two of these strips end to end.

(c) End to end, what lengths can you make with
   (i) three strips   (ii) four strips

(d) Which lengths between 1 cm and 20 cm are impossible to make with these strips?
(You can use a single strip on its own to make a length.)

(e) Design a set of five strips from which you can make lengths of 1 cm, 2 cm, 3 cm, ... as far up as you can go.

What is the longest length you can make?

B2 Another way to make lengths is to put strips side by side.

Here is a way to make 7 cm.

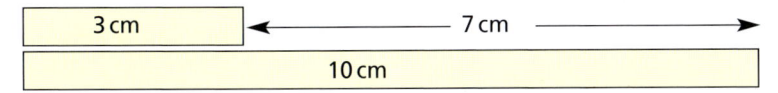

[3 cm]
[10 cm] ← 7 cm →

Here is a way to make 9 cm.

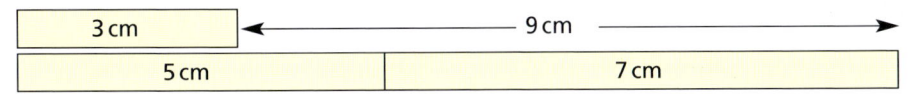

[3 cm]
[5 cm] [7 cm] ← 9 cm →

Using the strips at the top of the page, what lengths can you make by putting strips side by side?

B3 Design a set of four strips which can be used, either on their own, end to end or side by side, to make lengths of 1 cm, 2 cm, 3 cm, ... as far up as you can go.

What is the longest length you can make?

Review 4

1. Alex did an experiment about language learning. He gave people two tests.
   In test 1, people were given 20 words in an unfamiliar language and were asked to learn their meanings.
   In test 2, they were given 20 Chinese characters and asked to learn their meanings.
   In both cases they had three minutes in which to learn and were then tested.
   Here are Alex's results.

Person	A	B	C	D	E	F	G	H	I	J	K	L	M	N	O	P	Q	R
Score in test 1	9	5	12	10	4	7	13	15	9	8	11	10	12	10	6	8	7	11
Score in test 2	5	2	8	9	4	5	8	9	4	6	8	6	7	7	3	7	6	6

   Draw a scatter diagram and describe the correlation between the two test scores.

2. Solve these equations.
   (a) $4(n - 3) = n + 3$
   (b) $2(x - 1) = 6(x - 5)$

3. Two consecutive numbers multiply together to make 3422.
   Use trial and improvement to find the smaller of these numbers.

4. Work out each missing angle.

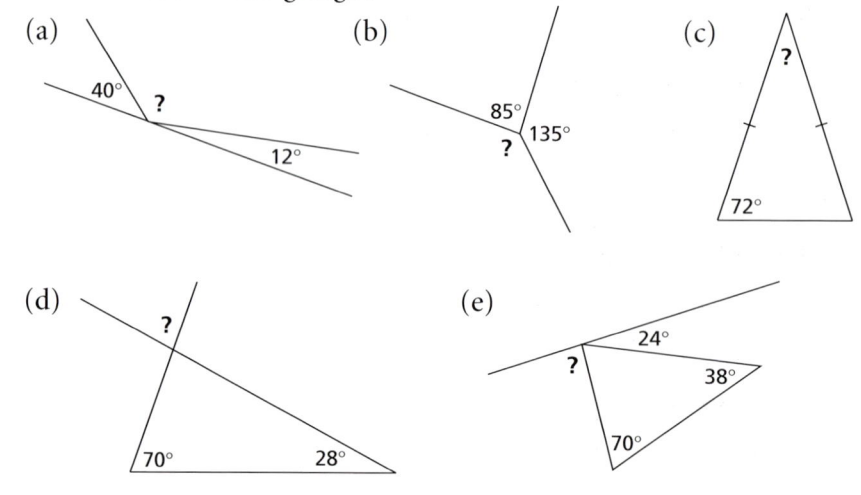

   (a) 40°, 12°, ?
   (b) 85°, 135°, ?
   (c) 72°, ?
   (d) 70°, 28°, ?
   (e) 24°, 38°, 70°, ?

5. John has tried to calculate 1.6×0.4 and has made a mistake.

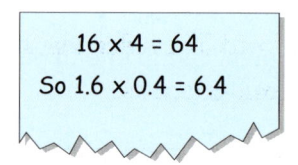

   16 × 4 = 64
   So 1.6 × 0.4 = 6.4

   What is the correct answer for 1.6×0.4?

This square tiling pattern appears in the Azem Palace in Damascus.

Some shapes have been labelled and some points and lines are shown to help describe any transformations.

Some angles are marked too.

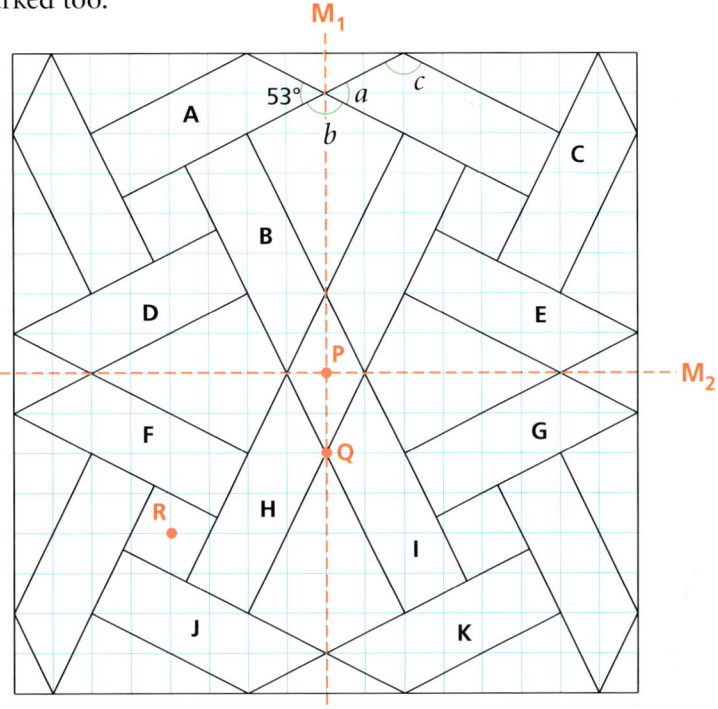

6 (a) Calculate the size of each angle a, b and c.
   (b) Which of these are a pair of corresponding angles?

7 Which shape is the image of these?
   (a) Shape J after the translation $\begin{bmatrix} 8 \\ 8 \end{bmatrix}$
   (b) Shape G after a rotation of 180° about P

8 (a) Describe a reflection that maps E to G.
   (b) Describe a translation that maps D to K.
   (c) Describe a rotation that maps J to F.

9 Describe fully a transformation that maps
   (a) J to K   (b) K to A   (c) H to F   (d) C to H

10 (a) What is the image of shape F after a reflection in M_1 followed by a reflection in M_2?
   (b) What single transformation could replace this pair of transformations?

11 (a) Find the image of I after a rotation of 180° about P followed by a reflection in M_2.
   (b) What single transformation could replace this pair of transformations?

12 Find the missing angle in each of these.

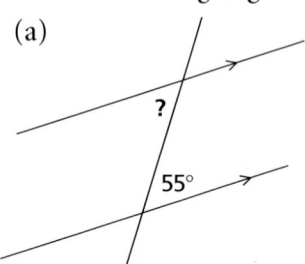

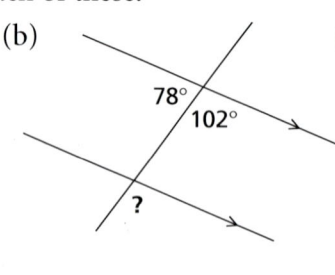

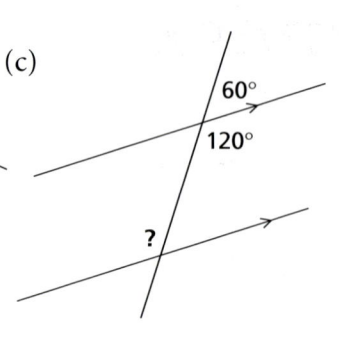

(a) ?, 55°

(b) 78°, 102°, ?

(c) 60°, 120°, ?

13 Helen has 5 bags of sweets and 2 extra sweets.

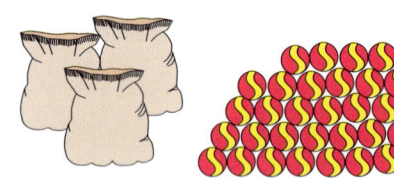

Bryony has 3 bags of sweets and 44 extra sweets.

Suppose that there are n sweets in each bag.

(a) Write an expression for the number of sweets Helen has.

(b) Write an expression for the number of Bryony's sweets.

(c) They each have the same number of sweets.
Write an equation and solve it to find how many sweets there are in a bag.

14 Use trial and improvement to find a solution (correct to 1 d.p.) of the equation
$$x^2 + x = 40$$

You could use a table like this to record your results.

x	$x^2 + x$	Too small/too big
4		
5		

15 Solve each of these equations.

(a) $3f + 2 = 47 - 2f$

(b) $44 - 3g = 20 - g$

(c) $50 - 4x = 2$

(d) $50 - 8h = 47 - 6h$

(e) $5y - 4 = 26 - y$

(f) $6k - 17 = 1 - 3k$

16 Find the missing angle.

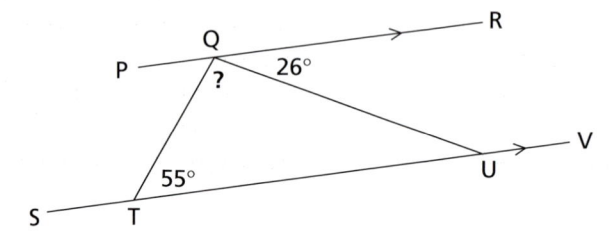

Index

addition and subtraction
   of decimals 115–116
   of fractions 20–21
algebraic expressions
   multiplying out such as $3(2n + 1)$, $n(n - 3)$ and corresponding factorisation 13–17
   simplifying, adding and subtracting algebraic fractions 17–18
angles
   of a polygon 91–97
   around a point, on a line, of a triangle, vertically opposite 151–153
   formed by parallel lines 153–155

circle
   area of 127–131
   radius, diameter and circumference of 46–52, 129–130
correlation 138–142
corresponding angles 154–155
cross section of solid 4–9

data, effective presentation of 99–102
decimals
   adding and subtracting 115–116
   multiplying and dividing 117–121
   recurring 124
   square, cube, square root and cube root of 118
differences in linear sequence 108
direct proportion 85–89
distance–time graph 32–33
division
   of decimals 120–121
   of fractions 25–26

enlargement by ray method 55–59
equation
   forming to find unknown value in proportion problem 88–89
   graph of linear 64–71
   solving after substituting into formula 38
   solving quadratic with no linear term 150
   solving simple linear 37, 144
   solving with subtracted unknown 144–150

exact value of π 51
experiment, use of correlation in 142
formula
   changing the subject of 39–43
   for nth term of linear sequence 108–113
   substituting into to obtain an equation 38
fractions
   equivalence, adding and subtracting 20–21
   multiplying and dividing 21–26
frequency chart 101
function, graph of linear 64–71

graph
   distance–time 32–33
   of linear function 64–71
   of real-life function of time 27–35
   of time series 99
   speed–time 33–35

implicit form for linear graph 71
iterative solution of equation 168–173

line of best fit 139–140
linear equation
   graph of 64–71
   solving 37, 144
linear sequence
   finding formula for nth term 110–113
   substituting into formula for nth term 108–109
   term-to-term rule for 107
loci 73

median, use in quartering method to detect correlation 140–141
misleading statistical charts 103–106
multiplication
   of fractions 21–24
   of decimals 117–120

parallel lines, angles formed by 153–155
path from a set of instructions 97
percentage
   finding an increase and decrease as 78–81
   increase and decrease using decimal multiplier 77
   to compare proportions 81–82

π 51
plane of symmetry 11
polygon, angles of 91–97
prism, volume and cross section of 9–11
proportion 85–89

quartering method to detect correlation 140–141

ratio as a single number to test for direct proportion 85–88
ray method for enlargement 55–59
recurring decimal 124
reflection 158–159, 161–165
reflection symmetry of three-dimensional object 11
regular polygon 95–97
rotation 159–162, 164–165

scale drawing 73–76
scale factor for enlargement 55–59
scatter diagram 135–142
solid, cross section of 4–9
speed–time graph 33–35

spreadsheet for trial and improvement 172–173
square, cube, square root and cube root of decimal 118
statistical charts, misleading 103–106
statistical survey
    effective presentation of data from 99–102
    use of correlation in 142
stem-and-leaf table 101
symmetry, reflection, of three-dimensional object 11

time series graph 99
transformations of points and shapes 156–165
translation 156–158, 161–164
travel graph 32–33
trial and improvement 167–173
triangle, construction given right angle, hypotenuse and side 75
two-way table 100

vector 156–158
vertically opposite angles 151–155
volume of prism 10–11